ERROR CODING FOR
ARITHMETIC PROCESSORS

ELECTRICAL SCIENCE
A Series of Monographs and Texts

Editors: Henry G. Booker and Nicholas DeClaris
A complete list of titles in this series appears at the end of this volume

ERROR CODING FOR ARITHMETIC PROCESSORS

T. R. N. Rao

Department of Electrical Engineering
University of Maryland, College Park

ACADEMIC PRESS   *New York and London · 1974*

A Subsidiary of Harcourt Brace Jovanovich, Publishers

ACADEMIC PRESS, INC.
111 Fifth Avenue, New York, New York 10003

United Kingdom Edition published by
ACADEMIC PRESS, INC. (LONDON) LTD.
24/28 Oval Road, London NW1

Library of Congress Cataloging in Publication Data

Rao, Thammavarapu R   N   Date
    Error coding for arithmetic processors.

    (Electrical science)
    Includes bibliographical references.
    1.   Error-correcting codes (Information theory)
2.   Computer arithmetic and logic units.   I.   Title.
QA268.R36 1974   519.4   73-22381
ISBN 0–12–580750–3

To

Rajyalaxmi

Ramakrishna, Chandrika, and Radhika

"విబుధ జనుల వలన విన్నంత కన్నంత
తెలియవచ్చినంత తేటపరుతు"

బమ్మెర పోతరాజు

What I have heard or seen
from many a scholar,
a pious hope it has been
to render it crystal clear.

Translated from Bammera Pothanna's "Bhagavatham."

Contents

Foreword

Computation without error remains an illusive goal of considerable importance in certain critical applications which require sophisticated and extensive computation with a high degree of system reliability. Recent advances in solid state technology have provided individual devices with exceptional reliability. In some systems, this improvement in device reliability has obtained sufficient systems reliability. However, in others, the large number of devices required has negated the improvement in reliability at the systems level. Such problems can be solved by the unlikely development of a perfect device which never fails. In the absence of such a device, one can expect greater use of the techniques of fault-tolerant computing to obtain improved systems reliability. Such improvement is not obtained without cost in performance or equipment, but in some applications, the cost is justifiable.

The technology of fault-tolerant computing is, at the present time, in its infancy and much development is still needed. In particular, more research and development directed toward realizable implementations is needed. There exists a substantial amount of literature on the subject, but the number of successful applications remains limited. However, the need for fault-tolerant computing will probably increase rather than decrease in the future.

Professor Rao, in this book, considers arithmetically invariant codes, which is an important technique of fault-tolerant computing. He has surveyed the extensive literature on the subject and organized it in a form which should be extremely useful to researchers in this area.

HARVEY L. GARNER
Moore School of Electrical Engineering
University of Pennsylvania
Philadelphia, Pennsylvania

Preface

Since the early work of Diamond (1955), there has been continuous progress in arithmetic coding theory. This field derives its strength and vitality from the classical works of Brown (1960), Peterson (1961), Chien (1964–1972), Massey (1964), Garner (1966), and others.

The purpose of this book is to combine the available knowledge on arithmetic codes and bring it under one cover for the benefit of students and researchers in this field of specialization. The presentation here includes the necessary mathematical background and error control preliminaries in order for this book to serve as a viable text in an advanced undergraduate or a graduate level course. The preliminary drafts of this book have been used for a graduate course in arithmetic coding and are presently being used in a new (modified version of the above) course entitled "Arithmetic Codes and Fault-Tolerant Computing" at the University of Maryland.

The first two chapters provide the student with a minimal mathematical background in algebra, number theory, and error control techniques. Simple mathematical models for registers, arithmetic processors, and elementary arithmetic operations are introduced. Chapter 3 introduces arithmetic codes, definitions, and code classifications. Single-error detection using AN codes and separate codes is also presented. Chapters 4 and 5 cover single-error-correcting codes using

AN codes and separate codes, respectively. A section on byte-error correction using higher radix codes is included in Chapter 4. In Chapter 6 we introduce code conversion algorithms leading to a presentation of the large distance codes, namely, Barrows–Mandelbaum codes and Chien–Hong–Preparata codes. Chapter 7 begins with the systematic nonseparate category of codes and covers codes for burst errors and iterative errors. We conclude the book with a presentation of codes and their applications in Chapter 8.

I truly believe that I was very fortunate in having the opportunity to study under Professor Harvey L. Garner at the University of Michigan. I have benefited from and been profoundly influenced by his works, thoughts, and teachings. Many fundamental definitions and notations and much terminology used in this book reflect this background and influence.

This book contains a number of sections which report on the results of our research on "Residue Codes and Application to Arithmetic Processors" during the years 1967–1972. This research was made possible by grants from the National Science Foundation and the National Aeronautics and Space Administration. I acknowledge very gratefully the initial guidance and valuable suggestions on the subject matter of this book given by Professor Robert T. Chien of the University of Illinois. The comments and suggestions from Professor Oscar N. Garcia of South Florida University and Dr. Se June Hong of the IBM Corporation on portions of this book have been most helpful.

Finally I wish to thank my wife, Rajyalaxmi, for her understanding nature and constant encouragement without which this book would not have been possible.

T. R. N. RAO

1 INTRODUCTION AND BACKGROUND

This chapter introduces the minimal mathematical background that will be useful in understanding subsequent chapters. In the first section we introduce algebraic structures such as *groups*, *subgroups*, *factor groups*, *rings*, and *fields*. In Section 1.2 the concepts of linear congruences, residues, and their properties are discussed. We also discuss here some elementary theory of numbers and the Chinese Remainder Theorem. Finally, in Section 1.3, the concept of a register and its number representation systems are introduced. The register operations, i.e., the elementary arithmetic operations on a register, are mathematically derived in that section.

1.1 ALGEBRAIC STRUCTURES

Groups $\{G; *\}$

A group G is a set of objects or elements for which an operation is defined and for which the following axioms (G1–G4) hold. Let $a, b, c, \ldots$ be the elements of the group and let $*$ denote the operation in the group. We may write $a * b = c$, $a * c = f$, etc. If the operation is addition, then we call it an additive group and use the symbol $+$. If the operation is multiplication, we call it a multiplicative group and use the symbol $\cdot$ instead of $*$.

AXIOM G1   (Closure)

For any two elements $a, b \in G$, $a * b$ must be an element of G.

AXIOM G2   (Associative law)

For any $a, b, c \in G$, $a * (b * c) = (a * b) * c$.

The associativity axiom implies that the order of performing the operation is immaterial and therefore the parentheses may be dropped.

AXIOM G3   (Identity)

There is an identity element e in the group G such that for any $a \in G$, $a * e = e * a = a$. (If the operation is addition, then 0 denotes the identity element and $a + 0 = 0 + a = a$. If the group is multiplicative, then 1 is the identity element and $a \cdot 1 = 1 \cdot a = a$ for any $a \in G$.)

AXIOM G4   (Inverse)

Every element of the group must have an inverse element with respect to the identity. That is, for any $a \in G$, there must be another element $a' \in G$ such that $a * a' = a' * a = e$. (Once again for additive groups, the inverse of a is denoted by $-a$, and $a + (-a) = 0$, and for multiplicative groups the inverse of a is denoted by a^{-1}, and $a \cdot a^{-1} = 1$.)

In addition to the four laws just mentioned, a group may satisfy the commutative law (G5) given below. If it does, the group is called an Abelian group or a commutative group.

AXIOM G5   (Commutativity)

For any $a, b \in G$, $a * b = b * a$.

LEMMA 1.1

In any group $\{G; +\}$:

if $a + b = a + c$, then $b = c$ (left cancellation);
if $a + b = c + b$, then $a = c$ (right cancellation).

Proof

Let $a + b = a + c$. From G4 there exists $-a \in G$. Adding that on the left to both sides and applying G2 we obtain $(-a + a) + b = (-a + a) + c$. Therefore $0 + b = 0 + c$ and $b = c$, as was to be proved. By a similar argument, we prove right cancellation.

LEMMA 1.2

In any group $\{G; *\}$, the identity element is unique.

Let there be two identity elements e_1 and e_2 and

$$a = a * e_1 = a = a * e_2$$

By the left cancellation law we have $e_1 = e_2$, as was to be proved.

EXAMPLES OF GROUPS

(1)   The set of all integers under the operation of addition is an Abelian group as can be easily verified. We denote the group by $\{Z; +\}$.

(2)   The set of all real numbers under addition is an Abelian group denoted as $\{R; +\}$.

(3)   The set of all real numbers, excluding zero, is an Abelian group under multiplication denoted as $\{R'; \cdot\}$.

(4)   The set of all nonsingular $n \times n$ matrices over reals is a (non-Abelian) group under matrix multiplication.

(5)   The set of integers $Z_4 = \{0, 1, 2, 3\}$ under operation $\oplus$, which is addition modulo 4, is an Abelian group, and is denoted as $\{Z_4; \oplus\}$.

The number of elements of a group is said to be the *order* of the group. If the order of a group is small, then its structure can be described conveniently by an operation table such as the one shown in Table 1.1 for $\{Z_4, \oplus\}$.

Table 1.1

$\oplus$	0	1	2	3
0	0	1	2	3
1	1	2	3	0
2	2	3	0	1
3	3	0	1	2

Subgroups and factor groups

A subset of elements H of a group G is called a *subgroup* if it satisfies all the axioms of a group. The operation of the subgroup H is, of course, the same as the group G. In order to determine whether a subset H of a group G is a subgroup, the following two axioms (S1 and S2) only need to be satisfied. The other axioms follow naturally from these two.

AXIOM S1

For all $a, b \in H$, $a * b \in H$ (closure).

AXIOM S2

For any $a \in H$, a' exists in H (inverse).

Since the elements of H are also elements of G, associativity holds in H just as in G. Also from S1 and S2 identity exists in H, thus completing the proof that H is a group by itself.

EXAMPLES

For the group $\{Z_4; \oplus\}$ described by Table 1.1 there are exactly three subgroups.

(1) The group G is a subgroup of itself.

(2) The elements $\{0, 2\}$ form a subgroup as can be easily verified.

(3) The element $\{0\}$ alone satisfies all the axioms of a group and therefore is a subgroup. This is a trivial subgroup.

A subgroup H of G is a *proper subgroup* if $H \neq \{e\}$ and $H \neq G$. Example (2) is the only proper subgroup of $\{Z_4; \oplus\}$.

Consider the set $kZ = \{kn \,|\, n \in Z\}$ for a constant integer $k \in Z$. kZ is the set of all integer multiples of the fixed integer k. For example,

$$3Z = \{0, -3, 3, -6, 6, -9, 9, \ldots\}$$
$$8Z = \{0, -8, 8, -16, 16, \ldots\}$$

We can show that kZ is a subgroup of $\{Z, +\}$ for any $k \in Z$. First observe that kZ is closed under addition

$$kl + km = k(l + m) \in kZ$$

Next $k0 = 0 \in kZ$. Finally, for each $kl \in kZ$, there exists $-kl \in kZ$. Therefore kZ is a subgroup of $\{Z; +\}$. It can be shown that every subgroup of $\{Z; +\}$ is of the form kZ. (See, for example, Johnson [1, p. 35].)

Let G be an Abelian group under multiplication, and let H be a subgroup of G. We define a coset gH of H

$$gH = \{gx \,|\, \text{for all } x \in H\}$$

for each $g \in G$. For an additive group G and its subgroup H, the cosets of H are denoted by $g_1 + H, g_2 + H, \ldots,$ for $g_1, g_2, \ldots \in G$. Consider the group $\{Z; +\}$ and its subgroup $H = 3Z$. Consider two cosets of H, namely $1 + 3Z, -5 + 3Z$.

$$1 + 3Z = \{1, -2, 4, -5, 7, -8, 10, \ldots\}$$
$$-5 + 3Z = \{-5, -8, -2, -11, 1, -14, \ldots\}$$

It is easily observed that the two cosets above are the same, and that -5 belongs to the coset $1 + 3Z$ and 1 belongs to the coset $-5 + 3Z$. In fact, two elements belong to the same coset if their difference is in H. This can be stated as follows.

THEOREM 1.3†

For an (additive) Abelian group G and its subgroup H, either $a + H = b + H$ or $(a + H) \cap (b + H) = \varnothing$. Further, $a + H = b + H$ iff $b - a \in H$.

The proof is left as an exercise to the student. ($\cap$ denotes intersection and $\varnothing$ denotes the null set.) All the cosets of the subgroup $5Z$ can be written as

$$\{5Z, 1 + 5Z, 2 + 5Z, 3 + 5Z, 4 + 5Z\}$$

We represent the set (of cosets) above by the notation $Z/5Z$.

If G is an additive Abelian group and H is a subgroup, then there is a natural way to define an operation in G/H (the set of cosets of H in G) which is due to the following:

$$(a + H) + (b + H) = (a + b) + H \qquad \text{for all} \qquad a, b \in G$$

THEOREM 1.4

If G is an (additive) Abelian group, and H is a subgroup of G, then $\{G/H, +\}$, with addition as defined above, is an Abelian group.

Outline of proof

For any $a, b, \in G$, $(a + H) + (b + H) = a + b + H \in G/H$, and therefore G/H is closed under $+$. Since

$$((a + H) + (b + H)) + (c + H) = (a + H) + ((b + H) + (c + H))$$
$$= (a + b + c) + H$$

---

† The theorems stated here for additive groups can also be stated for multiplicative groups, but the cosets of subgroups have to be written as aH, bH, etc. [1, 2].

associativity holds. $(0 + H) = H \in G/H$ such that for any $a + H \in G/H$, $(a + H) + H = H + (a + H) = a + H$, giving us identity. Also

$$(a + H) + (b + H) = (a + b) + H = (b + a) + H = (b + H) + (a + H)$$

Finally, $(-a + H) = -(a + H)$ for each $a + H$.   Q.E.D.

$\{G/H; +\}$ above is called the *factor group* or *quotient group* of G by H. The quotient group of $\{Z; +\}$ by the subgroup kz will be denoted by Z_k and

$$Z_k = Z/kZ = \{kZ, 1 + kZ, 2 + kZ, \ldots, k - 1 + kZ\}$$

For simplicity, the elements of Z_k can be denoted by $\{0, 1, 2, \ldots, k - 1\}$ and the addition in Z_k is then addition modulo k. As examples in Z_5, $3 + 3 = 1$, $2 + 4 = 1$, $4 + 1 = 0$, and so on.

EXAMPLE

Consider the Abelian group $G = \{Z_{15}; +\}$ where $+$ denotes addition modulo 15. Consider the subset $H = \{0, 3, 6, 9, 12\}$. Under addition modulo 15, H is a group and is therefore a proper subgroup of G.

Forming cosets of H as follows we obtain:

$$H = \{0, 3, 6, 9, 12\}$$
$$1 + H = \{1, 4, 7, 10, 13\}$$
$$2 + H = \{2, 5, 8, 11, 14\}$$

These three cosets $\{H, 1 + H, 2 + H\}$ exhaust G. G/H is the factor group containing these three elements, namely H, $1 + H$, $2 + H$, satisfying

$+$	H	$1 + H$	$2 + H$
H	H	$1 + H$	$2 + H$
$1 + H$	$1 + H$	$2 + H$	H
$2 + H$	$2 + H$	H	$1 + H$

This group G/H has the same structure as the Z_3 under addition modulo 3.

Rings and fields

The group we discussed previously has only one operation. We may define two operations, addition and multiplication, on the elements of the sets we have considered so far. But they are not Abelian groups with respect to each operation. For instance, the real numbers are an Abelian group under addition but not under multiplication. However, R', the real numbers excluding zero, is a group under multiplication but not under addition. So likewise $\{Z; +\}$ is an Abelian group, but $\{Z; \cdot\}$ is not. The algebraic system of a set R under addition and multiplication is called a *ring* if R1–R4 hold in R.

AXIOM R1

$\{R; +\}$ is an Abelian group.

AXIOM R2

R is closed under multiplication.

AXIOM R3

R is associative under multiplication.

AXIOM R4

Multiplication is distributive with respect to addition in R. That is, for any $a, b, c \in R$,

$$a \cdot (b + c) = a \cdot b + a \cdot c$$

and

$$(a + b) \cdot c = a \cdot c + b \cdot c$$

A ring $\{R; +; \cdot\}$ is a *commutative ring* if R5 holds in R.

AXIOM R5

$\forall\, a, b \in R, a \cdot b = b \cdot a.$

Further, a ring $\{R;\, +;\, \cdot\}$ may (but not necessarily) satisfy the following.

AXIOM R6

There exists a unit (or multiplicative identity), denoted 1 in R, such that for any $a \in R, a \cdot 1 = 1 \cdot a = a.$

AXIOM R7

For each $a \in R$, $a \neq 0$, there exists an element $a^{-1} \in R$ such that $a \cdot a^{-1} = a^{-1} \cdot a = 1.$

A ring $\{R;\, +;\, \cdot\}$ satisfying R5–R7 is called a *field*.

EXAMPLES

(1) $\{Z;\, +;\, \cdot\}$ is a commutative ring having a unit element 1. Note that it does not satisfy R7 and therefore it is not a field.

(2) $\{Z_k;\, +;\, \cdot\}$, where $+$ and $\cdot$ are addition modulo k and multiplication modulo k, respectively, is a commutative ring with unit. It can be shown that this ring is a field if k is a prime.

THEOREM 1.5

In any ring $\{R;\, +;\, \cdot\}$, $a, b \in R$,

$$a \cdot 0 = 0 \cdot a = 0$$

and

$$a \cdot (-b) = (-a) \cdot b = -(a \cdot b)$$

Proof

$(a \cdot 0) + 0 = a \cdot 0 = a \cdot (0 + 0) = (a \cdot 0) + (a \cdot 0)$. By left cancellation we get $0 = a \cdot 0$. Similarly, we can get $0 \cdot a = 0$. Since

$$0 = a \cdot 0 = a \cdot (b + (-b)) = a \cdot b + a \cdot (-b)$$

and R is an Abelian group under addition, $a \cdot (-b) = -(a \cdot b)$. By a similar argument we can show that $(-a \cdot b) = -(a \cdot b)$, and this completes the proof.

Ideals and residue classes

A subset S of a ring R is an ideal if axioms I1 and I2 hold.

AXIOM I1

S is an additive subgroup of R.

AXIOM I2

For any $a \in S$ and any $r \in R$, $a \cdot r \in S$.

For the ring $\{Z; +; \cdot\}$, consider the subset of Z consisting of all multiples of 5. For any two multiples of 5, their sum is also a multiple of 5. For any multiple of 5, its multiplicative product with an integer is a multiple of 5. Thus the set of all multiples of 5 in Z is an ideal. Similarly for any $k \in Z$, the set kZ is an ideal of $\{Z; +; \cdot\}$; conversely an ideal in Z consists of all multiples of some integer.

In the theory of rings, ideals play a role similar to the role of subgroups in Abelian groups. Since an ideal S is also a subgroup, cosets can be formed. In the case of rings, the cosets are called residue classes. All the distinct elements of an ideal S of R are placed in the first row of a rectangular array with the 0 element at the left. Then any element not in the ideal can be chosen as leader of the first residue class, and the

rest of the residue class is formed by adding the leader to each element
of the ideal:

$$S = \{a_1 = 0 \qquad a_2 \qquad a_3 \qquad a_4 \qquad \cdots\}$$
$$\{r_1\} = \{r_1 \qquad r_1 + a_2 \quad r_1 + a_3 \quad r_1 + a_4 \quad \cdots\}$$
$$\{r_2\} = \{r_2 \qquad r_2 + a_2 \quad r_2 + a_3 \quad r_2 + a_4 \quad \cdots\}$$
$$\vdots$$

The leading element of each row is, as before, a previously unused
element. It is to be noted that all theorems of cosets apply also to residue
classes. Addition and multiplication of the residue classes is defined:

$$\{r_1\} + \{r_2\} = \{r_1 + r_2\}$$
$$\{r_1\} \cdot \{r_2\} = \{r_1 r_2\},$$

where $\{r_i\}$ defines the residue class containing r_i. The set of residue
classes under these two operations satisfies all axioms of a ring and
therefore is called the ring of residue classes. We study further the ring
of residue classes in the integers in the next section under the heading of
linear congruences and residues.

1.2 THEORY OF DIVISIBILITY AND CONGRUENCES

We study here some elementary number theory as well as theory
relating to the properties of integers. By integers we mean not only
the natural numbers $\{1, 2, 3, \ldots\}$ but also zero and the negative integers
$\{-1, -2, -3, \ldots\}$.

The sum, difference, and product of two integers a and b are also
integers, but the quotient resulting from the division of a by b (for
$b \neq 0$) may be an integer or a noninteger. If $a/b = c$ for some integer c,
then we say b *divides* a, or b *is a divisor of a*, and denote this by $b|a$.
Further, we have the following:

THEOREM 1.6   (Euclidean division)

Every integer a is uniquely representable in terms of a positive integer b in the form

$$a = bq + r \qquad \text{for} \quad 0 \le r < b \tag{1.1}$$

r is the least nonnegative remainder obtained by division of an integer a by a positive integer b and has a special notation:

$$r = |a|_b \tag{1.2}$$

The greatest common divisor

In what follows, we consider only the positive divisors of numbers. An integer that divides each one of the integers $a, b, \ldots, m$ is said to be a *common divisor* of them. The greatest common divisor is abbreviated gcd and written as $\gcd(a, b, \ldots, m)$, or simply $(a, b, \ldots, m)$.

If $\gcd(a, b) = 1$, then a, b are said to be *relatively prime*. It can be shown that any common divisor t of a and b also divides $\gcd(a, b)$.

EXAMPLE

2, 6, 12 are common divisors of $\{24, 60\}$ but $(24, 60) = 12$. $(6, 10, 15) = 1$ and therefore 6, 10, 15 are relatively prime, but not pairwise prime. (To be pairwise prime every pair of the set must be relatively prime; 7, 9, 13 are pairwise prime.)

Euclidean algorithm

To obtain the greatest common divisor of two integers a and b, the Euclidean division algorithm is employed as follows:

$$a = bq_1 + r_1, \qquad 0 \le r_1 < b$$

Every integer that divides a and b also divides r_1. Similarly every integer that divides b and r_1 also divides a. From the above, $(a, b) = (b, r_1)$.

$$b = r_1 q_2 + r_2 \qquad 0 \le r_2 < r_1$$
$$r_1 = r_2 q_3 + r_3 \qquad 0 \le r_3 < r_2$$
$$\vdots$$
$$r_{n-2} = r_{n-1} q_n + r_n \qquad 0 \le r_n < r_{n-1}$$
$$r_{n-1} = r_n q_{n+1} \qquad r_{n+1} = 0$$

This algorithm terminates when some $r_{n+1} = 0$. This termination must occur in a finite number of steps, since the sequence $r_1, r_2, r_3, \ldots$ is a decreasing sequence of positive integers and a and b are finite integers. Also

$$(a, b) = (b, r_1) = (r_1, r_2) = \cdots = (r_{n-1}, r_n) = r_n$$

Therefore (a, b) is given by the smallest positive integer in the sequence $r_1, r_2, \ldots, r_n$. Consider as an example 1017 and 279. We write

$$1017 = 279 \times 3 + 180$$
$$279 = 180 \times 1 + 99$$
$$180 = 99 \times 1 + 81$$
$$99 = 81 \times 1 + 18$$
$$81 = 18 \times 4 + 9$$
$$18 = 9 \times 2 + 0$$

giving us $(1017, 279) = 9$. From the above

$$9 = 81 - 18 \times 4$$
$$= 81 - (99 - 81)4$$
$$= 5 \times 81 - 4 \times 99$$
$$= 5(180 - 99) - 4 \times 99$$
$$= 5 \times 180 - 9 \times 99$$
$$= -9 \times 279 + 14 \times 180$$
$$= -9 \times 279 + 14(1017 - 279 \times 3)$$
$$9 = 14 \times 1017 - 51 \times 279$$

This, in general terms, means that if $(a, b) = d$, then d can always be expressed as

$$d = ar + bs$$

for some two integers r and s. The above can be used to prove the following.

THEOREM 1.7

If $(a, b) = d$, then $(a/d, b/d) = 1$.

Proof

Assume $(a/d, b/d) = t$. Then t divides both a/d and b/d or td divides both a and b. That is, td is a common divisor of a and b, and therefore must divide d, which is impossible unless $t = 1$, as was to be proved.

THEOREM 1.8

If a divides bc and $(a, b) = 1$, then a divides c.

Proof

If $(a, b) = 1$, there are integers r, s such that $ar + bs = 1$. Multiplying the latter by c, we have $acr + bcs = c$. Since a divides ac and bc, a must also divide c.

THEOREM 1.9   (Unique factorization theorem)

Every integer $a > 1$ can be expressed as a product of one or more primes. Further, this expression as a product of primes is unique up to the order of the factors.

The proof is omitted here but is readily available in many texts on number theory (see LeVeque [3] or Vinogradov [4]).

The least common multiple

An integer that is a multiple of each one of the integers from the set $a, b, \ldots, m$ is said to be their *common multiple*. The smallest positive common multiple is called the *least common multiple* (lcm), and is written $\mathrm{lcm}\langle a, b, \ldots, m \rangle$ or simply $\langle a, b, \ldots, m \rangle$.

THEOREM 1.10

If a divides m and b divides m, then $\langle a, b \rangle$ divides m. Further,

$$\langle a, b \rangle = |ab|/d \tag{1.3}$$

Outline of Proof

Let $a \mid m$ and $b \mid m$; then m is a common multiple of a, b. Since m is a multiple of a, $m = ak$ for some integer k. But m is also a multiple of b and hence ak/b must also be an integer. Setting $(a, b) = d$, $a = a_1 d$, and $b = b_1 d$, we have $ak/b = a_1 dk/b_1 d = a_1 k/b_1$, where $(a_1, b_1) = 1$. Therefore b_1 divides k and we can write $k = b_1 t = bt/d$, where t is any integer. Hence we have

$$m = (ab/d)t$$

Conversely, every m of this form is a multiple of a as well as of b, and therefore this form gives all the common multiples of a, b. The smallest positive one of these multiples is the least common multiple and is obtained by setting $t = 1$ and taking the absolute value. Therefore

$$\langle a, b \rangle = |ab|/d \qquad \text{Q.E.D.}$$

Linear congruences and residues

The congruence notation, first introduced by Gauss, is a convenient form of expressing the fact that two integers a and b differ by a multiple of a positive integer m. We say that a *is congruent to* b *modulo* m and write

$$a \equiv b \pmod{m}$$

Here m is a divisor of $a - b$, and we can also write $a = b + km$ for some integer k. Clearly, we have the following.

THEOREM 1.11

$a \equiv a \pmod{m}$ (reflexive).

If $a \equiv b \pmod{m}$, then $b \equiv a \pmod{m}$ (symmetric).

If $a \equiv b \pmod{m}$ and $b \equiv c \pmod{m}$, then $a \equiv c \pmod{m}$ (transitive).

Proof

Since m divides $a - a = 0$, $a \equiv a \pmod{m}$. If m divides $a - b$, m divides $b - a$ and therefore we have the symmetric property. If $a \equiv b \pmod{m}$, $b \equiv c \pmod{m}$, then $a = b + mk_1$ and $b = c + mk_2$ for some k_1 and k_2. Therefore $a = c + mk_2 + mk_1 = c + m(k_1 + k_2)$, and hence $a \equiv c \pmod{m}$, proving transitivity.   Q.E.D.

Congruence, by virtue of the above, is an equivalence relation on the set of integers.

EXAMPLES

$$61 \equiv -2 \pmod{7}, \qquad 4^2 \equiv 3 \pmod{13}$$

Other properties of congruence are given by the following.

THEOREM 1.12

Given $a \equiv \alpha \pmod{m}$ and $b \equiv \beta \pmod{m}$, we have

$$a + b \equiv \alpha + \beta \pmod{m}$$
$$a - b \equiv \alpha - \beta \pmod{m} \qquad (1.4)$$
$$a \cdot b \equiv \alpha \cdot \beta \pmod{m}$$

Proof

From the hypothesis we have that m divides $a - \alpha$ and $b - \beta$. Therefore m divides $(a - \alpha) + (b - \beta) = (a + b) - (\alpha + \beta)$, which clearly means that $a + b \equiv \alpha + \beta \pmod{m}$. Also m divides $(a - \alpha) - (b - \beta)$. This means that m divides $(a - b) - (\alpha - \beta)$. In other words, $a - b \equiv \alpha - \beta \pmod{m}$. From the hypothesis, we have $a = \alpha + k_1 m$, $b = \beta + k_2 m$ for some k_1 and k_2. Therefore

$$a \cdot b = (\alpha + k_1 m)(\beta + k_2 m) = \alpha\beta + (\alpha k_2 + \beta k_1 + k_1 k_2 m)m = \alpha\beta + k_3 m$$

and this proves the last part.

Since $k \equiv k \pmod{m}$, we obtain the following trivially. from Theorem 1.12.

COROLLARY 1.13

Given $a \equiv b \pmod{m}$, we have the following:

$$a + k \equiv b + k \pmod{m}$$
$$a - k \equiv b - k \pmod{m} \qquad (1.5)$$
$$ak \equiv bk \pmod{m}$$

for any k.

This corollary shows that we can add, subtract, or multiply both sides of a congruence by a constant. But it is not always legitimate to cancel a factor from a congruence. For example,

$$25 \equiv 5 \pmod{10}$$

By cancellation of the factor 5 from the numbers 25 and 5, we get the false result $5 \equiv 1 \pmod{10}$. The reason is simple. The first congruence states that $25 - 5$ is a multiple of 10. The cancellation of a factor from a congruence is legitimate if the factor is relatively prime to the modulus m. That is, $ka \equiv kb \pmod{m}$ states that m divides $ka - kb = k(a - b)$. If m, k are relatively prime, then m must divide $a - b$, or $a \equiv b \pmod{m}$, and cancellation holds. In general terms, the following can be stated.

THEOREM 1.14

Given $(k, m) = d$, we have $ka \equiv kb \pmod{m}$ iff $a \equiv b \pmod{(m/d)}$.

Proof

Let $a \equiv b \pmod{(m/d)}$. Then m/d divides $a - b$, which means that m divides $d(a - b)$. Since $(k, m) = d$, k/d is an integer and m must also divide $(k/d)d(a - b) = k(a - b)$. This gives us $ka \equiv kb \pmod{m}$.

Conversely, let $ka \equiv kb \pmod{m}$. This means that m divides $ka - kb = k(a - b)$, and m/d divides $(k/d)(a - b)$. From Theorem 1.7, $(m/d, k/d) = 1$, and from Theorem 1.8 we have that m/d must divide $a - b$. This gives us $a \equiv b \pmod{(m/d)}$, and completes the proof.

It is obvious that any integer x is congruent modulo m to exactly one element of the set

$$0, 1, 2, \ldots, \quad m - 1$$

For, if x is congruent to two distinct elements y and w of this set, then y and w are congruent to each other and m divides their difference $y - w$, which is impossible. For the same reason, given any integer x and positive integer m it can be written uniquely as

$$x = qm + r, \qquad 0 \le r < m$$

Using the notation previously given,

$$r = |x|_m$$

and r is called the *least nonnegative residue* of x modulo m.

An integer that is congruent to x is also called the residue of x mod m. The infinite set of all such integers congruent to x modulo m is called a congruence class or *residue class* of x modulo m. There are clearly m residue classes of integers modulo m, each of them containing exactly one element from $\{0, 1, 2, \ldots, m - 1\}$.

A complete set of residues is a set consisting of exactly one element from each of the residue classes of integers modulo m, and is also called a *complete residue system*. In other words, a complete system of residues is any m integers no two of which are congruent to one another. $\{0, 1, 2, \ldots, m - 1\}$ and $\{1, 2, \ldots, m\}$ are examples of complete residue systems of integers modulo m. The elements of a complete residue system need not be consecutive integers, and for $m = 5$, the set $\{10, -4, -3, 3, 9\}$ is a complete residue system modulo 5. A complete residue system modulo m can be characterized as a set $\{a_1, a_2, \ldots, a_m\}$ with the following properties:

(a) If $i \neq j$, then $a_i \not\equiv a_j \pmod{m}$.

(b) If x is any integer, there exist some a_i $(1 \leq i \leq m)$ for which $x \equiv a_i \pmod{m}$.

The complete residue system modulo m, $\{0, 1, \ldots, m - 1\}$, is used here most frequently. We also call this the set of integers modulo m and denote it by Z_m.

There is another kind called the *reduced* (or *incomplete*) residue system modulo m. This system consists of integers $a_1, a_2, \ldots, a_n$ incongruent modulo m, such that if x is any integer relatively prime to m, then there exist some a_i $(1 \leq i \leq n)$ for which $x \equiv a_i \pmod{m}$. In other words, a reduced residue system is a set of representatives, one from each of the residue classes, containing only the integers relatively prime to m. The reduced residue system modulo m is denoted by $G(m)$ hereafter.

As an example, $G(8) = \{1, 3, 5, 7\}$ is a reduced residue system modulo 8 and consists of those integers of Z_m relatively prime to 8. The number of elements in a reduced residue system modulo m is denoted by $\phi(m)$, and is called the *Euler's function of m*. In the ring $\{Z_m; +; \cdot\}$, where $+$ and $\cdot$ are addition and multiplication modulo m, respectively, the reduced residue system, $G(m)$ is an Abelian group under multiplication modulo m.

Linear congruences

Consider a linear congruence in one unknown x

$$ax \equiv b \pmod{m} \tag{1.6}$$

This has a unique solution modulo m provided that a is relatively prime to m. The proof of the above is as follows. If $(a, m) = 1$, there exist integers p, q such that $ap + mq = 1$ or $ap \equiv 1 \pmod{m}$. Multiplying by b we get the congruence $apb \equiv b \pmod{m}$. This gives us $apb \equiv ax \pmod{m}$, and by cancellation of a, which is valid for $(a, m) = 1$, we obtain $pb \equiv x \pmod{m}$ or $x = |pb|_m$, the desired solution.

In the result above, p is a multiplicative inverse of a in integers modulo m, and dividing the congruence by a is equivalent to multiplying by p. This we can write as

$$x = \left|\frac{b}{a}\right|_m = |bp|_m$$

Now consider the congruence (1.6) when $(a, m) = d > 1$. This means that $ax - mk = b$ for some k. Since d divides both a and m it must divide b. In other words, the congruence is not solvable unless d divides b. Therefore assume d divides b. Writing $a = a'd$ and $b = b'd$ we obtain

$$a'dx \equiv b'd \pmod{m} \tag{1.7}$$

From Theorem 1.14, and from the fact that $(d, m) = d$, the congruence (1.7) is equivalent to

$$a'x \equiv b' \bmod\left(\frac{m}{d}\right) \tag{1.8}$$

Further, from Theorem 1.7, $(a/d, m/d) = (a', m/d) = 1$, and therefore (1.8) is solvable and has a unique solution for x modulo m/d. If x_0 is such a solution, we have

$$a'x_0 \equiv b' \pmod{m/d} \qquad \text{and} \qquad x_0 = \left|\frac{b'}{a'}\right|_{m/d}$$

and x_0 is also a solution of (1.6).

Noting that $0 \leq x_0 < m/d$, consider $y = x_0 + (m/d)t$ as a generalized solution of (1.6). Then

$$ay = ax_0 + \frac{m}{d}a \cdot t = ax_0 + ma't \equiv ax_0 \pmod{m}$$

and therefore y is indeed a solution of (1.6). The distinct values of y in the range $0 \leq y < m$ are exactly d of them, namely

$$x_0, \quad x_0 + \frac{m}{d}, \quad x_0 + \frac{2m}{d}, \quad \ldots, \quad x_0 + \frac{(d-1)}{d} m$$

This gives us the following important result.

THEOREM 1.15

Given $(a, m) = d$, the linear congruence

$$ax \equiv b \,(\text{mod } m)$$

is solvable for x in Z_m if d divides b; and if it is solvable, there are exactly d solutions for x in Z_m.

EXAMPLE

$$9x \equiv 3 \,(\text{mod } 21)$$

$\gcd(9, 21) = 3$, and dividing by 3

$$3x \equiv 1 \,(\text{mod } 7)$$

$\gcd(3, 7) = 1$ and therefore it has a unique solution in Z_7, namely

$$x_0 = \left| \frac{1}{3} \right|_7 = 5$$

Besides 5, $5 + 7 = 12$, and $5 + 2 \times 7 = 19$ are all the solutions for the original congruence.

Consider the two congruences in one unknown

$$x \equiv a \,(\text{mod } m_1), \qquad x \equiv b \,(\text{mod } m_2) \tag{1.9}$$

Then $x = a + m_1 y \equiv b \,(\text{mod } m_2)$ or $m_1 y = b - a \,(\text{mod } m_2)$.

From Theorem 1.15, the last congruence is solvable only if (m_1, m_2)

divides $b - a$. From Theorem 1.14 we can divide this congruence by $(m_1, m_2) = d$, and obtain

$$\frac{m_1}{d} y \equiv \frac{(b - a)}{d} \left(\text{mod } \frac{m_2}{d}\right) \tag{1.10}$$

The congruence (1.10) has a unique solution for y modulo m_2/d and $x = a + m_1 y$ has a unique solution modulo $m_1 m_2/d$. Also y has exactly d solutions modulo m_2 and therefore x will have the same number of solutions modulo $m_1 m_2$. This proves the following.

LEMMA 1.16

The two congruences in one unknown,

$$x \equiv a \,(\text{mod } m_1), \qquad x \equiv b \,(\text{mod } m_2)$$

are solvable only if $d = (m_1, m_2)$ divides $a - b$. If they are solvable, then there is a unique solution modulo $\langle m_1, m_2 \rangle$, and exactly d solutions modulo $m_1 m_2$.

Lemma 1.16 can be generalized to any arbitrary number of moduli and stated as Theorem 1.17. Its proof is omitted here but is available in many textbooks on number theory (see LeVeque [3] or Vinogradov [4]).

THEOREM 1.17

A necessary and sufficient condition that the set of congruences $x \equiv a_i \,(\text{mod } m_i)$ $(i = 1, 2, \ldots, n)$ be solvable is that for every pair of moduli m_i, m_j, (m_i, m_j) divide $a_i - a_j$. The solution, if it exists, is unique modulo $\langle m_1, m_2, \ldots, m_n \rangle$.

As a special case, when the moduli are pairwise relatively prime, i.e., for every pair m_i, m_j, $(m_i, m_j) = 1$, then $\langle m_1, m_2, \ldots, m_n \rangle = m_1 m_2 \cdots m_n$. Hence we have

THEOREM 1.18 (Chinese Remainder Theorem)

The set of congruences $x \equiv a_i \pmod{m_i}$ $(i = 1, 2, \ldots, n)$, in which the moduli $m_1, m_2, \ldots, m_n$ are pairwise relatively prime, is solvable, the solution being unique modulo the product $m_1 m_2 \cdots m_n$.

EXAMPLE

From Theorem 1.18 the following congruences

$$x \equiv 1 \pmod{3} \tag{1.11}$$

$$x \equiv 3 \pmod{4} \tag{1.12}$$

$$x \equiv 4 \pmod{5} \tag{1.13}$$

have a unique solution modulo $3 \cdot 4 \cdot 5 = 60$. To find that solution we solve the first two congruences, which gives a unique solution modulo $3 \cdot 4 = 12$, and use this with the third congruence to obtain the answer. From (1.11), $x = 3y + 1$, and substituting this into (1.12)

$$3y + 1 \equiv 3 \pmod{4}$$
$$3y \equiv 2 \pmod{4}$$

or

$$y = \left| \frac{2}{3} \right|_4 = |2 \cdot 3|_4 = 2$$

Therefore $x = 3y + 1 = 3 \cdot 2 + 1 = 7$ in integers modulo 12 or

$$x \equiv 7 \pmod{12} \tag{1.14}$$

The next step is to solve (1.14) and (1.13) simultaneously. From (1.14) $x = 7 + 12w$ for some w, and substituting into (1.13)

$$x = 7 + 12w \equiv 4 \pmod{5}$$
$$12w \equiv 4 - 7 \equiv 2 \pmod{5}$$
$$2w \equiv 2 \pmod{5}$$

Since $(2, 5) = 1$, we can cancel 2 from the above, giving $w = 1$, and $x = 7 + 12w = 19$.

Table 1.2 illustrates the one-to-one correspondence between $x \in Z_{30}$ and the residues of x with respect to the moduli 2, 3, and 5. This system of representation is called the *residue number system* (RNS) or modular arithmetic [5, 6].

Table 1.2 RNS representing Z_{30}

x	$|x|_2$	$|x|_3$	$|x|_5$	x	$|x|_2$	$|x|_3$	$|x|_5$
0	0	0	0	15	1	0	0
1	1	1	1	16	0	1	1
2	0	2	2	17	1	2	2
3	1	0	3	18	0	0	3
4	0	1	4	19	1	1	4
5	1	2	0	20	0	2	0
6	0	0	1	21	1	0	1
7	1	1	2	22	0	1	2
8	0	2	3	23	1	2	3
9	1	0	4	24	0	0	4
10	0	1	0	25	1	1	0
11	1	2	1	26	0	2	1
12	0	0	2	27	1	0	2
13	1	1	3	28	0	1	3
14	0	2	4	29	1	2	4

In general terms a residue number system uses n residues with respect to moduli m_1, m_2, ..., m_n which are pairwise prime, and represents nonredundantly (i.e., on a one-to-one basis) the set of integers modulo $M = m_1 \cdot m_2 \cdots m_n$. Let two numbers $x, y \in Z_M$ be represented as n residues (or n-tuples) as follows:

$$x = (x_1, x_2, \ldots, x_n), \qquad x_i = |x|_{m_i} \qquad (i = 1, 2, \ldots, n)$$
$$y = (y_1, y_2, \ldots, y_n), \qquad y_i = |y|_{m_i} \qquad (i = 1, 2, \ldots, n)$$

It follows from the theory of residues that the sum $S = |x + y|_M$ is given by

$$S = (|x_1 + y_1|_{m_1}, |x_2 + y_2|_{m_2}, \ldots, |x_n + y_n|_{m_n})$$

Also the product $P = |x \cdot y|_M$ is given by

$$P = (|x_1 \cdot y_1|_{m_1}, |x_2 \cdot y_2|_{m_2}, \ldots, |x_n \cdot y_n|_{m_n})$$

The corresponding residues are added or multiplied, and there are no carries between the residue positions. These properties give the advantage of fast addition and multiplication operations and have received considerable attention from computer designers and researchers [7, 8] over the last decade. The RNS computers suffer from a major disadvantage in sign determination [8], and therefore remain more or less theoretical machines. However, the residue error codes, which are derived from the same theory as RNS, appear very practical and more promising in the design of self-detecting–correcting processors, and comprise the major topic in this book.

1.3 REGISTERS AND NUMBER REPRESENTATION SYSTEMS

In this section we define precisely the concepts of register and its number representation systems.

A *register* **A** is an ordered set of n cells or n digits, and is denoted by an n-tuple $(A_{n-1}, A_{n-2}, \ldots, A_1, A_0)$, where each digit A_i can assume any value from the set $D = \{0, 1, 2, \ldots, r - 1\}$. The set D is called the *digit set* and r is called the *radix* (and sometimes *base* or *modulus*) of the digit position. If $r = 2$, the register is termed binary; $r = 3$, ternary; $r = 10$, decimal; and so on. For a binary register, each digit (in this case, bit) can be represented by a basic storage element, such as a flip-flop or a ferrite memory core. On the other hand, each digit in a decimal may require four or more such elements. Thus, a register can be associated with n sets of physical devices, one set of devices for each digit. Although the word register so often refers to physical devices or storage elements, we will use the earlier definition, which is that a register is an ordered set of n digits. Since the value assumed by a digit changes with time and the sequence of operations on a register, each digit is a function of time. Therefore we write $\mathbf{A}(t) = (A_{n-1}(t), A_{n-2}(t), \ldots, A_0(t))$ as a function whose domain is the time axis and whose range is the set of all n-tuples over D.

The integer value A of a register $\mathbf{A}$, denoted by $\delta_I(\mathbf{A})$, is given by the polynomial

$$A = \delta_I(\mathbf{A}) = \sum_{i=0}^{n-1} A_i r^i \tag{1.15}\dagger$$

The integer value of $\mathbf{A}$ so defined has a range $0 \le \delta_I(\mathbf{A}) \le r^n - 1$. Alternatively, δ_I is a map of all n-tuples over D onto the set of non-negative integers $\{0, 1, 2, \ldots, r^n - 1\}$. We denote the latter set by Z_{r^n} or Z_m, where $m = r^n$. Since there are no negative integers in this representation system, the complement number systems, namely, the radix complement (RC) and diminished radix complement (DRC) systems, are introduced. For specific cases, when $r = 2$, the radix complement corresponds to 2's complement, and the diminished radix complement corresponds to 1's complement. Further, if $r = 10$, the RC system is called the 10's complement system, and DRC the 9's complement.

In complement systems, a negative integer $-N$ is represented by $\overline{N}$, the complement of N with respect to r^n or $r^n - 1$. That is, $\overline{N} = r^n - N$ (in RC systems) or $\overline{N} = r^n - 1 - N$ (in DRC systems) represents $-N$. If δ_{rI} and $\delta_{r-1, I}$ represent the maps for RC, and DRC systems, respectively, we write

$$\delta_{rI}(\mathbf{A}) = \begin{cases} \delta_I(\mathbf{A}) & \text{if } \delta_I(\mathbf{A}) < r^n/2 \\ \delta_I(\mathbf{A}) - r^n & \text{if } \delta_I(\mathbf{A}) \ge r^n/2 \end{cases} \tag{1.16}$$

$$\delta_{r-1, I}(\mathbf{A}) = \begin{cases} \delta_I(\mathbf{A}) & \text{if } \delta_I(\mathbf{A}) < r^n/2 \\ \delta_I(\mathbf{A}) - (r^n - 1) & \text{if } \delta_I(\mathbf{A}) \ge r^n/2 \end{cases} \tag{1.17}$$

By letting m denote r^n or $r^n - 1$, as the case may be, and letting $S(\mathbf{A})$ denote the sign function of $\mathbf{A}$ such that

$$S(\mathbf{A}) = \begin{cases} 0 & \text{for } \delta_I(\mathbf{A}) < r^n/2 \\ 1 & \text{for } \delta_I(\mathbf{A}) \ge r^n/2 \end{cases} \tag{1.18}$$

† Here registers are represented by boldface letters, such as $\mathbf{A}$, and their integer values are given in italic.

we can combine Equations (1.16) and (1.17) as

$$\delta_{rI}(\mathbf{A}) = \delta_I(\mathbf{A}) - S(\mathbf{A}) \cdot m \qquad \text{(for} \quad m = r^n)$$
$$\delta_{r-1,I}(\mathbf{A}) = \delta_I(\mathbf{A}) - S(\mathbf{A}) \cdot m \qquad \text{(for} \quad m = r^n - 1) \qquad (1.19)$$

$S(\mathbf{A})$ is the sign function of $\mathbf{A}$ and is zero [denoting positive $\delta_{rI}(\mathbf{A})$ or $\delta_{r-1,I}(\mathbf{A})$] iff $\delta_I(\mathbf{A}) < r^n/2$. Whenever the radix r is even, the sign $S(\mathbf{A})$ is obtained by the most significant digit of $\mathbf{A}$, namely, A_{n-1}. If $A_{n-1} < r/2$, then $\delta_I(\mathbf{A}) < r^n/2$ and therefore $S(\mathbf{A}) = 0$ and $\delta_{rI}(\mathbf{A})$ and $\delta_{r-1,I}(\mathbf{A})$ are both nonnegative. This property of $S(\mathbf{A})$ can be observed from Table 1.3 for the radices $r = 2$ and $r = 10$. Due to this fact, A_{n-1} is called the sign bit in binary complement systems. Also for even $r > 2$, A_{n-1} contains the information regarding the sign implicitly along with the magnitude. On the other hand, when r is odd, $S(\mathbf{A})$, as defined by (1.18), is not given by A_{n-1} alone but is a function of all the A_i.

Table 1.3a Number representation systems, $r = 2$, $n = 4$

$\mathbf{A} = (A_3 A_2 A_1 A_0)$	$\delta_I(\mathbf{A})$	$\delta_{1I}(\mathbf{A})$	$\delta_{2I}(\mathbf{A})$	$S(\mathbf{A}) = A_{n-1}$
0 0 0 0	0	0	0	0
0 0 0 1	1	1	1	0
0 0 1 0	2	2	2	0
0 0 1 1	3	3	3	0
0 1 0 0	4	4	4	0
0 1 0 1	5	5	5	0
0 1 1 0	6	6	6	0
0 1 1 1	7	7	7	0
1 0 0 0	8	−7	−8	1
1 0 0 1	9	−6	−7	1
1 0 1 0	10	−5	−6	1
1 0 1 1	11	−4	−5	1
1 1 0 0	12	−3	−5	1
1 1 0 1	13	−2	−3	1
1 1 1 0	14	−1	−2	1
1 1 1 1	15	0	−1	1

Table 1.3b $r = 10, n = 3$

$\mathbf{A} = (A_2 A_1 A_0)$	$\delta_I(\mathbf{A})$	$\delta_{9I}(\mathbf{A})$	$\delta_{10I}(\mathbf{A})$	$S(\mathbf{A})$
0 0 0	0	0	0	0
0 0 1	1	1	1	0
0 0 2	2	2	2	0
⋮	⋮	⋮	⋮	⋮
4 9 9	499	499	499	0
5 0 0	500	−499	−500	1
5 0 1	501	−498	−499	1
⋮	⋮	⋮	⋮	⋮
9 9 8	998	−1	−2	1
9 9 9	999	0	−1	1

For the specific case when $\mathbf{A} = (r - 1, r - 1, \ldots, r - 1)$, we have $\delta_I(\mathbf{A}) = r^n - 1$ and $S(\mathbf{A}) = 1$, and from Equation (1.17) we have $\delta_{r-1,\,I}(\mathbf{A}) = r^n - 1 - (r^n - 1) = 0$, and therefore $\mathbf{A}$ represents a zero. Thus, there are two distinct representations for zero in DRC systems, and consequently there is one less integer represented in them. Also from the formulas (1.19) we get

$$-r^n/2 \le \delta_{rI}(\mathbf{A}) \le r^n/2 - 1 \tag{1.20}$$

$$-(r^n/2 - 1) \le \delta_{r-1,\,I}(\mathbf{A}) \le r^n/2 - 1 \tag{1.21}$$

If the number of distinct integers represented in a system is called the *range* of the system and is denoted by the symbol m, then

$$m = \begin{cases} r^n & \text{for integer representation systems} \quad (\delta_I) \\ r^n & \text{for RC systems} \quad (\delta_{rI}) \\ r^n - 1 & \text{for DRC systems} \quad (\delta_{r-1,\,I}) \end{cases} \tag{1.22}$$

Using the congruence notation given in Section 1.2, Equations (1.19) can be written as

$$\delta_{rI}(\mathbf{A}) \equiv \delta_I(\mathbf{A}) \,(\text{modulo } m) \qquad \text{for} \quad m = r^n$$

$$\delta_{r-1,\,I}(\mathbf{A}) \equiv \delta_I(\mathbf{A}) \,(\text{modulo } m) \qquad \text{for} \quad m = r^n - 1$$

or

$$|\delta_{rI}(\mathbf{A})|_m = |\delta_I(\mathbf{A})|_m = \delta_I(\mathbf{A}) = A \qquad (1.23)$$

and

$$|\delta_{r-1,\,I}(\mathbf{A})|_m = |\delta_I(\mathbf{A})|_m \qquad (1.24)$$

The equations above reveal that the RC and DRC representations have the same structure as the integers modulo m where m denotes r^n or $r^n - 1$, respectively. The reader may find it useful to verify Equations (1.23) and (1.24) from the examples of Table 1.3.

A model for elementary operations

For the operations below, we consider a simple model for the arithmetic processor. Let the processor at time t contain one internal operand A (in register $\mathbf{A}$), *and let* operand B, and an opcode Φ be applied as inputs at time t. The output of the arithmetic processor is given by R, which also represents the contents of register $\mathbf{A}$ at time $t + 1$.

$$R = \Phi(A, B) = \delta_I(\mathbf{A}(t + 1)) = A(t + 1)$$

R is a function of none, one, or both operands, depending on the opcode Φ. For more details and an illustration of this model the reader should see Section 2.1 (Figure 2.3).

Elementary operations

We discussed earlier the number representation systems for a general radix r. We consider here binary registers and arithmetic operations on these registers. The previously obtained equations can be rewritten for a binary register $\mathbf{A} = (A_{n-1}, A_{n-2}, \ldots, A_0)$ as follows:

$$\delta_I(\mathbf{A}) = \sum_{i=0}^{n-1} A_i 2^i = A$$

$$\delta_{1I}(\mathbf{A}) = \delta_I(\mathbf{A}) - S(\mathbf{A}) \cdot m = \sum_{i=0}^{n-1} A_i 2^i - A_{n-1}m = A - A_{n-1}m$$

$$\delta_{2I}(\mathbf{A}) = \sum_{i=0}^{n-1} A_i 2^i - A_{n-1}2^n = \sum_{i=0}^{n-1} A_i 2^i - A_{n-1} \cdot m = A - A_{n-1}m$$

where $m = 2^n - 1$ for 1's complement case, and $m = 2^n$ for 2's complement case. We will show here that in addition to the integer representation structure the addition operation structures of complement systems correspond to the addition in the ring of integers modulo m.

Binary addition (Φ_1: ADD)

There are two types of binary addition structures. They correspond to both 1's and 2's complement additions. In a 2's complement addition, the carry generated from the most significant stage is discarded, whereas in a 1's complement addition such a carry is passed around and is applied to the least significant stage. This end-around-carry (EAC) is shown in Figure 1.1 by the dashed line and is required only for 1's

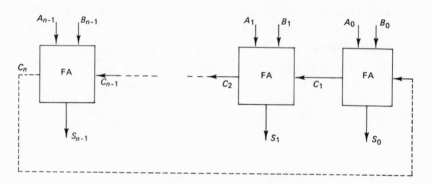

Figure 1.1 A parallel binary (ripple-carry) adder.

complement addition. Also FA denotes a full adder module and c_i denotes a carry into the ith module.

Since the carry from the most significant stage, denoted c_n, has a value 2^n, a discard of such a carry amounts to subtracting 2^n from the integer sum of the operands. If $S = (S_{n-1}, \ldots, S_0)$ denotes the sum obtained as a result of 2's complement addition of the two n-tuples A and B, then

$$S = \delta_I(S) = \delta_I(A) + \delta_I(B) - c_n 2^n = A + B - c_n 2^n$$

Therefore

$$S \equiv A + B \,(\text{mod } m)$$

Thus, the 2's complement addition structure corresponds to addition modulo m, where $m = 2^n$. On the other hand, an application of end-around-carry (if it occurs in 1's complement addition) amounts to a discard of a value $2^n - 1$ from the integer sum. Once again, if S denotes the 1's complement sum, then

$$S = \delta_I(\mathbf{S}) = \delta_I(\mathbf{A}) + \delta_I(\mathbf{B}) - c_n(2^n - 1) = A + B - c_n(2^n - 1)$$
$$\equiv A + B \,(\text{mod } m)$$

This states that the 1's complement addition structure corresponds to addition modulo m, where $m = 2^n - 1$. Combining these, we write

$$R = \Phi_1(A, B) = |A + B|_m \tag{1.25}$$

Subtraction (Φ_2 :   SUB)

A major advantage of the complement number systems is that subtraction is carried out by complementing the subtrahend, and adding it to the other operand, the minuend, thus eliminating the need for a separate subtractor functional block. Since the complementation is easily implemented, this scheme is very commonly used. In the 1's complement system, each bit of the subtrahend is reversed, 1 to 0 and 0 to 1, and is added with end-around-carry. In the 2's complement system, there will be no end-around-carry, but a 1's complement of the subtrahend is added with an initial carry of $c_0 = 1$ into the rightmost stage. This will accomplish 2's complement subtraction.

If $\bar{B}$ denotes the complement of B, then $A - B = A + \bar{B} - m$. Therefore we write

$$\Phi_2(A, B) = |A + \bar{B}|_m = |A - B|_m \tag{1.26}$$

Logical shift right (Φ_3 :   LSHR)

Let the value of the register $\mathbf{A}$ at time t be defined as $\delta_I(\mathbf{A}(t)) = A(t)$. Let Φ_3 be an operation on $\mathbf{A}$ which shifts its components one place to the right, while moving a 0 into the vacated leftmost position. In Figure 1.2, which illustrates this operation, the large boxes are binary cells or flip-flops, and the small boxes labeled T denote transfer gates which perform the shift at the command of the shift pulse applied to them.

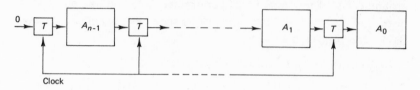

Figure 1.2 Logical shift right of register **A**.

If the operation is initiated at time t and is completed a unit time later, that is, at time $t + 1$, then we have the result R of the operation given by

$$R = \delta_I(\mathbf{A}(t + 1))$$

and

$$\mathbf{A}(t + 1) = (A_{n-1}(t + 1), \ldots, A_0(t + 1)) = \Phi_3(A(t))$$

We also have

$$A_{n-1}(t + 1) = 0$$
$$A_j(t + 1) = A_{j+1}(t) \qquad \text{for} \quad j = 0, 1, \ldots, n - 2$$

Hence we have

$$R = \delta_I(\mathbf{A}(t + 1)) = \sum_{i=0}^{n-1} A_i(t + 1)2^i$$

$$= \sum_{i=0}^{n-2} A_{i+1}(t)2^i = \sum_{i=0}^{n-2} A_{i+1}(t) \cdot 2^{i+1}/2$$

$$R = \tfrac{1}{2}\sum_{j=1}^{n-1} A_j(t)2^j + \tfrac{1}{2}A_0(t) - \tfrac{1}{2}A_0(t)$$

$$= \tfrac{1}{2}\sum_{j=0}^{n-1} A_j(t)2^j - \tfrac{1}{2}A_0(t)$$

$$= \tfrac{1}{2}(A(t) - A_0(t)),$$

which we write as

$$\Phi_3(A(t), B(t)) = \tfrac{1}{2}(A(t) - A_0(t)) \tag{1.27}$$

Arithmetic shift right (Φ_4 : ASHR)

This operation differs from the previous one in what enters the cell A_{n-1} during the shift. Instead of a 0 entering from the left into A_{n-1}, we now leave it unchanged. Consequently, if $\Phi_4(A(t)) = A(t + 1)$, then

$$A_{n-1}(t + 1) = A_{n-1}(t)$$

and

$$A_j(t + 1) = A_{j+1}(t) \qquad \text{for} \quad j = 0, 1, \ldots, n - 2$$

Calculating as before, we obtain

$$\Phi_4(A(t), B(t)) = A_{n-1}(t)2^{n-1} + \tfrac{1}{2}(A(t) - A_0(t)) \tag{1.28}$$

Cycle right (Φ_5 : CYR)

This operation also is different from the previous two operations in that quantity which enters the leftmost cell A_{n-1}. In this operation, A_0 enters the cell A_{n-1}. Hence, $A(t + 1)$ can be specified as

$$A_{n-1}(t + 1) = A_0(t)$$
$$A_j(t + 1) = A_{j+1}(t) \qquad \text{for} \quad j = 0, 1, \ldots, n - 2$$

Therefore

$$
\begin{aligned}
R = \delta_I(\mathbf{A}(t + 1)) &= \sum_{i=0}^{n-1} A_i(t + 1) \cdot 2^i \\
&= A_0(t) \cdot 2^{n-1} + \sum_{i=0}^{n-2} A_{i+1}(t) \cdot 2^i \\
&= A_0(t) \cdot 2^{n-1} + \tfrac{1}{2}(A(t) - A_0(t)) \\
&= (A_0(t)/2)(2^n - 1) + \tfrac{1}{2}A(t)
\end{aligned}
$$

We can therefore write

$$\Phi_5(A(t),\, B(t)) = \begin{cases} \left|\dfrac{A(t)}{2}\right|_m & \text{for} \quad \text{1's complement case} \\[2ex] \tfrac{1}{2}(A(t) + (m-1)A_0(t)) & \text{for} \quad \text{2's complement case} \end{cases}$$

(1.29)

Cycle left (Φ_6: CYL)

This operation is illustrated in Figure 1.3. $A(t+1)$ is specified by

$$A_0(t+1) = A_{n-1}(t)$$
$$A_j(t+1) = A_{j-1}(t) \qquad \text{for} \quad j = 1, 2, \ldots, n-1$$

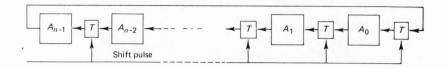

Figure 1.3   Cycle left of register **A**.

Hence

$$R = \delta_l(A(t+1)) = \sum_{i=0}^{n-1} A_i(t+1)2^i$$

$$= \sum_{i=1}^{n-1} A_{i-1}(t)2^i + A_{n-1}(t)$$

$$= \sum_{j=0}^{n-2} A_j(t)2^{j+1} + A_{n-1}(t)2^n - A_{n-1}(t)(2^n - 1)$$

$$= \sum_{j=0}^{n-1} 2A^j(t)2^j - A_{n-1}(t) \cdot (2^n - 1)$$

We therefore write

$$R = \Phi_6(A(t),\, B(t)) = |2A(t)|_m \qquad \text{for} \quad \text{1's complement case}$$

$$R = \begin{cases} 2A(t) - A_{n-1}(t) \cdot (2^n - 1) \\ |2A(t) + A_{n-1}(t)|_m & \text{for} \quad \text{2's complement case} \end{cases}$$

(1.30)

Table 1.4 Elementary operations

Elementary operation	Result $R = \Phi(A, B)$		Example $A = \delta_t(100101)^a = 37$, $B = \delta_t(011101) = 29$	
	1's compliment $m = 2^n - 1$	2's complement $m = 2^n$	1's complement $m = 63$	2's complement $m = 64$
Φ_1 (ADD)	$|A + B|_m$	$|A + B|_m$	$|37 + 29|_{63} = 3$	$|37 + 29|_{64} = 2$
Φ_2 (SUB)	$|A - B|_m$	$|A - B|_m$	$|37 - 29|_{63} = 8$	8
Φ_3 (LSHR)	$\frac{1}{2}(A - A_0)$	$\frac{1}{2}(A - A_0)$	$\frac{1}{2}(37 - 1) = 18$	18
Φ_4 (ASHR)	$\frac{1}{2}(A - A_0 + A_{n-1}2^n)$	$\frac{1}{2}(A - A_0 + A_{n-1}2^n)$	$\frac{1}{2}(37 - 1 + 64) = 50$	50
Φ_5 (CYR)	$\left|\frac{1}{2}A\right|_m$	$\frac{1}{2}(A + (m - 1)A_0)$	$\left|\dfrac{37}{2}\right|_{63} = 50$	$\dfrac{37 + 63}{2} = 50$
Φ_6 (CYL)	$|2A|_m$	$|2A + A_{n-1}|_m$	$|2 \times 37|_{63} = 11$	$|2 \times 37 + 1|_{64} = 11$
Φ_7 (SHL)	$|2A - A_{n-1}(t)|_m$	$|2A|_m$	$|2 \times 37 - 1|_{63} = 10$	$|2 \times 37|_{64} = 10$
Φ_8 (CLAD)	B	B	29	29

^{a} For simplicity A and B are used in place of $A(t)$ and $B(t)$.

Shift left (Φ_7: SHL)

This operation is the same as the one above, except that a 0 is moved into A_0. Hence

$$A_0(t + 1) = 0$$
$$A_j(t + 1) = A_{j-1}(t) \quad \text{for} \quad j = 1, 2, \ldots, n - 1$$
$$R = \delta_l(A(t + 1)) = 2A(t) - A_{n-1}(t) \cdot 2^n$$

We therefore write

$$\Phi_7(N_1, N_2) = \begin{cases} |2A(t)|_m & \text{for} \quad \text{2's complement case} \\ |2A(t) - A_{n-1}(t)|_m & \text{for} \quad \text{1's complement case} \end{cases} \quad (1.31)$$

Clear and add (Φ_8: CLAD)

$$\Phi_8(A(t), B(t)) = B(t) \quad (1.32)$$

Formulas (1.25)–(1.32) are listed in Table 1.4, where they are illustrated by means of an example.

PROBLEMS

1. Which of the axioms G1–G5 are satisfied by each of the following algebraic structures:

   (a) $\{Z^+; +\}$, Z^+ is the set of all positive integers.
   (b) $\{Z; -\}$.
   (c) $\{Q; \cdot\}$, Q is the set of rational numbers.
   (d) $\{N; +\}$, $N = \{0, 1, 2, \ldots\}$.
   (e) $\{Z; *\}$, where $a * b = a + b + a \cdot b$.
   (f) $\{R; *\}$, where $a * b = a^2 + b^2$.

2. Let $\{G; \cdot\}$ be a group. Prove that G is Abelian if either of the following holds:

   (a) $a^{-1} = a$ for all $a \in G$,   (b) $(a \cdot b)(a \cdot b) = a^2 \cdot b^2$.

3. Let $\{G; \cdot\}$ and $\{H; \cdot\}$ be two Abelian groups. $G \times H$ is the set of all ordered pairs (a, b) for $a \in G, b \in H$. Consider a binary operation $*$ on the elements of $G \times H$

$$(a_1, b_1) * (a_2, b_2) = (a_1 \cdot a_2, b_1 \cdot b_2)$$

Is $\{G \times H; *\}$ an Abelian group? Prove your answer.

4. Let $Z_k = \{0, 1, 2, \ldots, k - 1\}$ and $\oplus$, $\odot$ denote addition modulo k and multiplication modulo k, respectively. Show that the algebraic structure $\{Z_k; \oplus; \odot\}$ is a commutative ring. Is this structure a field? Explain.

5. In the ring Z_{15} obtain

   (a) *all* additive subgroups,
   (b) *all* multiplicative subgroups, and
   (c) *all* ideals.

6. Show that if $(x, y) = 1$, then $(x - y, x + y) = 1$ or 2.

7. Evaluate $d = (325, 4095)$ using the Euclidean division algorithm. Obtain d as a linear combination of 325 and 4095, that is, in the form $d = 325x + 4095y$.

8. Show that if $(a, b) = 1$, then $(ab, c) = (a, c) \cdot (b, c)$.

9. Obtain *all* the solutions for the following linear congruences. If there are no solutions, give the reason.

   (a) $10x \equiv 15 \pmod{25}$.
   (b) $21x \equiv 14 \pmod{35}$.
   (c) $6x \equiv 4 \pmod{9}$.

10. Solve the set of linear congruences

$$x \equiv 1 \pmod 2, \quad x \equiv 0 \pmod 3, \quad x \equiv 1 \pmod 5, \quad x \equiv 3 \pmod 7$$

REFERENCES

1. R. E. Johnson, "University Algebra." Prentice-Hall, Englewood Cliffs, New Jersey, 1966.
2. G. Birkhoff and S. MacLane, "A Survey of Modern Algebra." Macmillan, New York, 1960.

3. W. J. LeVeque, "Topics in Number Theory," Vol. 1. Addison-Wesley, Reading, Massachusetts, 1956.
4. I. M. Vinogradov, "Elements of Number Theory." Dover, New York, 1954.
5. H. L. Garner, The Residue Number System, *IRE Trans. Electron. Comput.* **EC-8**, 140–147 (June 1959).
6. "Modular Arithmetic Techniques," Tech. Documentary Rep. No. ASD-TDR-62-686. Lockheed Missiles and Space Co., Sunnyvale, California, January 1963.
7. N. S. Szabo and R. I. Tanaka, "Residue Arithmetic and its Applications to Computer Technology." McGraw-Hill, New York, 1967.
8. R. D. Merrill, Improving Digital Computer Performance Using Residue Number Theory, *IEEE Trans. Electron. Comput.*, **EC-13**, 93–101 (April 1964).

2 ARITHMETIC PROCESSORS AND ERROR CONTROL PRELIMINARIES

This chapter introduces the subject of arithmetic processors and error control techniques. In the first section, we define arithmetic processors (AP), arithmetic operations, and a model to describe their functional behavior. In the second section, *logic faults*, which are the cause of errors (both numerical and logical), are discussed. The origin and nature of logic faults are discussed. Then *arithmetic weight* of errors (or error numbers) is introduced. In Section 2.3, some error control techniques, namely, triple modular redundancy (TMR) as well as duplication and switching, are introduced.

2.1 ARITHMETIC PROCESSORS AND DIGITAL COMPUTERS

A digital computer is divided, for convenience, into a number of functional units (or subsystems), such as arithmetic processor (AP), control unit (CU), memory unit (MU), input/output unit (I/O unit), program unit (PU), etc. (see Figure 2.1). These divisions are useful from the point of view of computer organization and design. The various units are connected together appropriately to enable processing of the information required of it. Stated briefly, the function of the memory unit(s) is to store data, instructions, and intermediary results and to enable transfer of these to the arithmetic unit or I/O unit as required. The control unit handles the supervisory functions such as providing reference signals (or clock pulses) to initiate or terminate the various functions of these units. The I/O units provide the ability to communicate with the outside world. It may consist of tape readers or card readers and punch tape or punch card equipment. The arithmetic processor receives data (or operands) and operation commands (or instructions) from the memory and control units and performs the needed processing. The results of these operations are stored in the

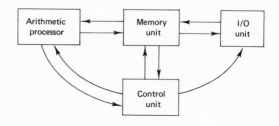

Figure 2.1   An organization of the functional units of a computer.

specified memory locations or retained in the arithmetic registers which are included as part of an AP. The nature and complexity of these various functional units depend on the class of problems they are to solve, and the instruction repertoire. Besides, the speed–cost trade-offs play a significant part in the makeup of each functional unit. No less important are factors such as the type of number system used, the arithmetic procedures (or algorithms) built in, the maintenance circuits (if any provided), and the required "reliability" or "dependability" of operation of the entire system. Therefore, any general characterization of any unit such as an AP will not be meaningful for any special-purpose computer operation but applies only in a rather general way. Since our interest here lies in (the error control coding for) arithmetic processors, we discuss their organization next.

Organization of an arithmetic processor

An arithmetic processor is the part of a computer that provides the logic circuits required to perform arithmetic operations such as ADD, SUBTRACT, MULTIPLY, DIVIDE, SQUARE-ROOT, etc. There are a number of other operations such as *complement, shift, rotate*, and *scale* which are classified at times as arithmetic operations and are performed by arithmetic units of large- and medium-sized computers. In order to perform these operations, an AP usually consists of a number of registers to store operands, intermediate results (partial sums or partial products), and logic blocks such as adders, division logic, overflow detection logic, shift–rotate logic, and in addition, several gating and decoding and control logic blocks.

A sample organization of a small arithmetic processor is shown in Figure 2.2. This setup shows four arithmetic registers, called *accumulator, addend, augend* (or *multiplicand*), and *multiplier–quotient* registers. Parallel adder, complement, and shift–rotate logic blocks are shown.

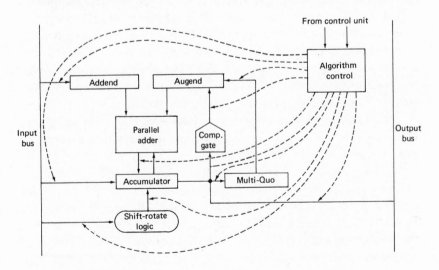

Figure 2.2 Sample organization of arithmetic processors.

Also indicated are an algorithm controller which receives the operation command from the control unit, and control signals to the logic blocks. These organizations come in a great variety and differ enormously in complexity. For the purpose of understanding the operation of an AP, we resort in the next section to a simple model which characterizes, in a rather general way, its operation from the point of view of error control coding logic. In this discussion one has to keep in mind the objective of this study, which is error control coding and not the design aspects of an AP. We discuss an organization only to enable a good understanding of the error control techniques presented later.

A simple model for arithmetic processors

An AP can be described by the inputs (operands and the operation command) and outputs (the sum or product, etc.). We can assume, without loss of generality, that one or more of the operands have already been supplied to the AP, and are stored in the appropriate registers; and the input required in such cases may be simply an operation command. " Shift right five times the contents of an accumulator " is an example of that type of instruction. For instructions such as ADD, generally one operand such as an augend is already available in the accumulator (or augend register), and another will be an input to the AP. Similarly, the outputs may simply be posted in one or more registers of the AP and can be retrieved from it on a separate operation command. Therefore, a block diagram of the type shown in Figure 2.3 can serve as a model for an AP.

Without any loss of generality, the input operand B, the internal operand A (available initially in the register $\mathbf{A}$), and the results R are each assumed to be of n binary digits (bits). The opcode for Φ may be k bits long, where k must be sufficiently large to accommodate all the possible operations of an AP. (Note that the number of different operation commands must be less than or equal to 2^k). Further, we could denote the results R by $R = \Phi(A, B)$. R is the result of the specified operation Φ (when the input operand is B and the internal operand is A).

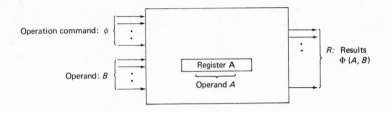

Figure 2.3   A model for an arithmetic processor.

The outputs R are numerical and/or logical results. The interpretation of the outputs as to the nature of their values is left to the control function. If Φ is the ADD instruction, then R may represent the sum, $A + B$ modulo m, denoted as $|A + B|_m$, where $m = 2^n$ for 2's complement binary logic, or $m = 2^n - 1$ for 1's complement binary logic.† If Φ denotes "cyclic shift left register **A** by one place," then R may represent the contents of register **A** after the execution of the instruction. If the register **A** represents the integer A (where $0 \leq A < 2^n$) before the cyclic shift, the integer value of **A** after the shift equals $|2A|_{2^n-1}$, which may also represent the numerical value of R. If Φ represents an operation, say "clear and add" (CLAD), then the register **A** will be replaced by the operand B. If Φ represents MULTIPLY, then R may represent the n most significant binary digits of the product of A and B. The least significant bits may be stored in one of the specified arithmetic registers.

The discussion above is intended as an overall view of the operations of the arithmetic units. A sample of operations has been given in Table 2.1. The seven operations listed first may be termed "elementary," and the rest "compound."

A compound operation can be performed by repeated use of one or more of the elementary operations; for example, MULTIPLY can be realized by repeated ADD, SHIFT operations. Therefore a number of medium-sized computers may not have any "built-in" compound

† For formulas which characterize these elementary operations and for examples the reader is advised to see Section 1.3 and in particular Table 1.4.

Table 2.1   Arithmetic operation commands

Elementary operations	
ADD	Φ_{ADD}
SUBTRACT (Complement and Add)	Φ_{SUB}
SHIFT RT^a	Φ_{SHR}
SHIFT LT	Φ_{SHL}
CYCLE LT	Φ_{CYL}
CLEAR and ADD	Φ_{CLAD}
STORE ACC	Φ_{STA}

Compound operations	
MULTIPLY	FLOATING PT. ADD
DIVIDE	SCALE
SQUARE-ROOT	COMPARE

^{a} We discussed previously in Chapter 1 two
types of SHIFT RT operations. One is " Logical
Shift Rt" and the other "Arith Shift Rt."

operations, but only a good set of elementary operations. Further, an
elementary operation may be assumed to use a logic block or logic
circuit just once. There are exceptions to this; for example, in the serial
ADD operation, a single full adder stage is repeatedly used. This aspect
of single-use or multiple-use of a logic block is an important concern
for error control coding and is therefore discussed in detail in Chapter 7
in the section on iterative errors.

Further, if an effective error-correcting scheme is available for all
elementary operations, then one could say that all arithmetic operations
can be error controlled. Then the elementary operations that are more
amenable to error control coding become an important part of our
study.

2.2   NATURE AND ORIGIN OF ERRORS IN AP's

The words *errors, logic faults, component failures, malfunctions,* and
troubles have been defined and used previously by different authors.
Unfortunately there is no uniformity in their definitions. Before we

attempt some definitions of our own here, it may be appropriate to make a clear distinction between the words component, logic element, logic network, functional unit, etc. (See Figure 2.4.)

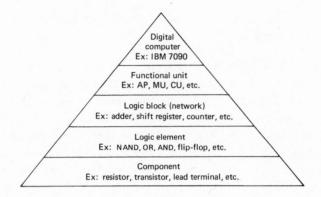

Figure 2.4 Subdivisions of a digital computer.

A *component* is the smallest building block of a digital computer, for example, a resistor, transistor, diode, or lead terminal. A *logic element* refers to a logic gate such as NAND, OR, AND, NOT, or a flip-flop. A *logic network* (or logic block) refers to a combinational network such as an adder, a complementer, an overflow detector, or a sequential network such as a shift register or a binary counter. Logic networks may vary in size considerably from one to another. A functional unit consists of a number of logic networks interconnected to perform any operation from a comprehensive set of operations. *Functional units* refer to large units such as an arithmetic processor, a control unit, or a memory unit.

Logic faults and their classifications

A *logic fault* or simply a *fault* is a deviation of logic variables from their specified values at one or more points of a logic network. Logic faults in an AP may be grouped broadly into two categories:

1.   Control logic faults, which are due to failures† or malfunctions in the control logic. As an example, a logic fault in the operation command decoder lets a wrong algorithm (Φ' instead of Φ) be executed.

2.   Arithmetic logic faults, which include logic faults in an adder, or division logic block, etc.

Both categories of faults relate to logic faults within the processor. If wrong inputs have been applied, that would constitute faults outside the premise of AP but are logic faults of the control processor or memory unit. Our concern is directed mainly toward obtaining error-free arithmetic processing through detection, location, and correction of errors originating in an AP. The techniques and codes studied for this purpose can also be used for error control of other functional units in an appropriate manner.

A logic fault was defined previously to be a deviation of logic variables from their specified values at one or more points of a logic network. A fault may invert a binary variable from 1 to a 0, from 0 to a 1, or force it to assume a constant logic value (such as stuck-at-1 or stuck-at-0). These faults are naturally caused by component failure(s) or malfunctions in a logic network. The origins of logic faults are related to component failures or malfunctions, electrical noise, overheating, etc., while the effects of logic faults are the errors in the outputs (or numerical results) of the AP.

The faults may be classified as *temporary* or *permanent*, *single* fault or *multiple* fault, *single-use* or *multiple-use*. This classification is based on the origin, nature, or use of the faulty network. *A temporary fault* is usually caused by a component malfunction or electromagnetic noise interference; its effects (the errors) cannot be reproduced under the program control. *A permanent fault is often caused by a component failure, and its effects are reproducible.* A *single fault* refers to one error-

---

† A component failure is the permanent destruction of a component such as a shorted or an open diode, or a grounded lead. A component malfunction is a temporary misbehavior due to outside noise interference, or overheating, or a marginal component.

inducing event, a *multiple fault* (double or triple, etc.) refers to several single faults occurring simultaneously. The effects of a multiple fault may be described as the cumulative effect of the individual single faults. A fault (either single or multiple) can be classified as *simple* (local) or *complex* (distributed) depending on its effect on the errors. A simple fault causes minimal damage on the outputs, or in other words, may produce an error of the type $\pm 2^j$; a complex fault causes extensive damage. Thus, a fault in a logic element, depending on the intricacy of the location of that logic element, is either simple or complex. By this definition, control faults are often classified as complex. This classification is based on the damage wrought on the results of an elementary operation or by the error value E, and more precisely, by its *arithmetic weight* $W(E)$, to be defined later.

The types of errors produced by a fault also differ from one operation to another. If an operation does not require the use of a faulty network, then the fault is virtually inactive for the duration of that operation. Another operation may use that faulty network just once, and still another operation may require repeated use of the same faulty network. Thus a logic fault will have different levels of damage on the output results of elementary operations and complex operations. This fact should be borne in mind by the error control coding designer. As an example, a *simple fault* may be due to a component failure in the carry generation logic of the ith stage of a parallel adder, so that the error generated (in the addition operation) has a value $E = \pm 2^j$. (The error word may have several successive nonzero positions, but we call the error a single error. A later section has precise definitions of arithmetic weights and single and double errors.) A *complex fault* in the same stage of an adder could be a cause for both the carry and sum to be in error. In the latter case, the error value $E \equiv \pm 2^j \pm 2^{j-1}$ (E here may be a single error or double error, depending on its error magnitude $|E|$). An important aspect to be noted here is that a fault becomes simple or complex depending on the role of the logic element that contains the fault or the failed component. Another point of interest is the need for preventing complex faults by suitable logic design. This objective can often conflict with the natural, time-old objective of the logic design, which is to realize logic circuits at minimal cost. An error control objective is to limit logic faults to simple and single-use categories, as far as possible, by design and by use of algorithms. This constraint is likely to limit the size of fanout of a logic element and increase the cost of hardware of the processor. This price, however, may be reasonable in view of the overall costs of alternative approaches such as the triple modular redundancy (TMR) or duplication and switching methods, which are described in Section 2.3.

Errors and arithmetic weight

Component failures or malfunctions produce logic faults. The logic faults generate erroneous results, or errors. Errors are thus the deviations in the outputs (numerical or logical) of the arithmetic processor. An error is said to occur in an operation Φ of AP whenever the actual output $\mathbf{R}' = (r'_{n-1}, r'_{n-2}, \ldots, r'_0)$ differs from the expected value $\mathbf{R} = (r_{n-1}, r_{n-2}, \ldots, r_0)$ specified by the designer. Therefore, the error word $\mathbf{E}$ (also called damage pattern or error pattern [1]) is

$$\mathbf{E} = (e_{n-1}, e_{n-2}, \ldots, e_0) = \mathbf{R}' - \mathbf{R} \qquad (2.1)$$

where $e_i = r'_i - r_i$ for $i = 0, 1, 2, \ldots, n - 1$. If we consider only binary outputs, r'_i and r_i can only be 0 or 1, and consequently e_i can be 0, 1, or -1.

EXAMPLE

Actual output

$$\mathbf{R}' = (110001), \qquad R' = \delta_I(\mathbf{R}')$$

specified output

$$\mathbf{R} = (101101), \qquad R = \delta_I(\mathbf{R})$$

error word

$$\mathbf{E} = (01\bar{1}\bar{1}00)$$

where $\bar{1}$ denotes -1 in the error word.

For each error word **E**, we define an error value E (or hereafter error) given by

$$E = \delta_I(\mathbf{E}) = \sum_{i=0}^{n-1} e_i 2^i \qquad (2.2)$$

and

$$E = R' - R \qquad (2.3)$$

For the example above, the error $E = \delta_I(\mathbf{E}) = \delta_I(01\bar{1}\bar{1}00) = 16 - 8 - 4 = 4$. This mapping δ_I of error words into errors is a conversion of the n-tuples into integer values, and is not a one-to-one correspondence. As examples, the error words $(01\bar{1}\bar{1}00)$, $(001\bar{1}00)$, and (000100) all correspond to the same error $E = 4$. The one most significant property of an error is its *arithmetic weight*, which is defined as follows.

The arithmetic weight† of an integer N, denoted $W(N)$ (in radix r), is defined as the *minimum* number of terms in an expression of the form

$$N = a_1 r^{j_1} + a_2 r^{j_2} + \cdots \qquad (2.4)$$

where $a_i \neq 0$, $|a_i| < r$ (see Peterson [2, Chapter 15]).

Since binary arithmetic processors and binary codes are of special interest, we may restate the definition above for binary systems. The binary arithmetic weight of N is the *minimum* number of terms in an expression of the form

$$N = a_1 2^{j_1} + a_2 2^{j_2} + \cdots \qquad (2.5)$$

where $a_i = 1$ or -1. For instance, the decimal number 31 has a binary representation 11111, but that is certainly not in a minimal form. Its minimal form is $10000\bar{1}$ ($\bar{1}$ denotes -1). Hence $W(31) = 2$, the number of terms in a minimal form. There may be more than one minimal form for a number N. For instance, $25 = 011001$ and $25 = 10\bar{1}001$. While the binary arithmetic weight of 31 is 2, its ternary arithmetic weight is 3, since $31 = 3^3 + 3 + 1$ is a minimal form. Thus the arithmetic weight of a number depends very much on the radix notation used. In order to simplify our notation, we observe the following. *Unless specifically stated otherwise, we will assume an arithmetic weight to refer to the binary arithmetic weight.*

† The arithmetic weight is also referred to as Peterson weight in some instances.

An important property of the arithmetic weight is that

$$W(N) = W(-N) \tag{2.6}$$

This can be very trivially observed from the expression (2.4) where each a_i can be negative or positive. If $N = \sum_{i=1}^{k} a_i r^{j_i}$, then $-N = \sum_{i=1}^{k} (-a_i) r^{j_i}$ and both N and $-N$ have exactly the same number of terms in their minimal forms. Another important property of the arithmetic weight is the triangular inequality given as

$$W(N_1 + N_2) \le W(N_1) + W(N_2) \tag{2.7}$$

This is clear if we consider addition of the numbers N_1 and N_2 which are initially in their minimal forms. Cancellation of nonzero terms or carries may occur, but the number of nonzero terms of the sum cannot possibly exceed the sum of the number of nonzero terms of N_1 and N_2. As said before, the minimal form (or minimal weight form) for N is not unique. As another example, a decimal number $19 = 2^4 + 2^2 - 1 = 2^4 + 2 + 1$. Reitwiesner [3] has shown that if an integer is given in such a form that the coefficients $a_i a_{i+1} = 0$ for $i = 0, 1, \ldots, n - 2$ in the expression

$$N = \sum_{i=0}^{n-1} a_i r^i, \qquad a_i = 0, \quad 1, \quad \text{or} \quad -1$$

then it is called the *nonadjacent form* (NAF) and it is also a minimal weight form. In Section 6.1 the interested reader can find further discussion on NAF and algorithms for conversion to these forms. An algorithm to determine by inspection the arithmetic weight of an integer expressed in binary form is available in the work of Garcia [4].

DEFINITION 2.1

Given integers N_1 and N_2, the arithmetic distance between N_1 and N_2, denoted $D(N_1, N_2)$, is given by $W(N_1 - N_2) = W(N_2 - N_1)$.

Let the specified (or correct) result and the actual (or possibly erroneous) result of an operation be N_1 and N_2, respectively. If the arithmetic distance between N_1 and N_2 is d, then a d-fold arithmetic error (or an error of weight d) is said to have occurred. If E in (2.3) is such that $W(E) = d$, then a d-fold arithmetic error E is said to have occurred. When two numbers are added, failure in one of the flip-flops of the accumulator or in one of the carry stages of the parallel adder may affect several consecutive bit positions in the sum due to carry propagation; the arithmetic weight and distance are explicitly defined to treat such errors as single errors.

Errors in finite ring arithmetic

Due to the finite size of the arithmetic registers, adders, and so on, the operands N_1 and N_2 are limited to a finite range of values. Let Z_m denote the finite ring of integers modulo m, namely $\{0, 1, \ldots, m - 1\}$.† The additive inverse of N in Z_m is often called the complement of N, and is denoted by $\overline{N} = m - N$ and

$$\overline{N} \equiv -N \,(\text{mod } m)$$

The arithmetic weight as defined earlier satisfies $W(N) = W(-N)$. However, for N and $\overline{N}$ in Z_m, $W(N)$ may not be equal to $W(\overline{N})$. As an example, consider Z_{63}, where $\overline{32} = 63 - 32 = 31$. We have

$$W(32) = 1, \qquad W(\overline{32}) = W(31) = 2$$

This is an undesirable feature, since error magnitude does not uniquely specify its arithmetic weight. A shorted inverter in the jth stage of an adder may generate errors of $+2^j$ and -2^j at different times, and therefore cause errors of different weight. To alleviate this difficulty, Rao and Garcia [5] have introduced modular arithmetic weights for numbers in a finite ring.

---

† Since we are concerned only with the addition operation in $Z_m = \{0, 1, \ldots, m - 1\}$, we need only to say that Z_m is an additive Abelian group. Algebraic ring properties will be of interest in consideration of both addition and multiplication operations.

DEFINITION 2.2

The modular weight of an integer $N \in Z_m$, denoted by $W_m(N)$, is given by

$$W_m(N) = \min(W(N), W(\overline{N}))$$

In the binary representation for $31 \in Z_{63}$, we have the following:

$$W(31) = 2$$
$$\overline{31} = 63 - 31 = 32, \qquad W(32) = 1$$

Therefore, $W_m(31) = W_m(32) = 1$.

Similarly, for N_1, $N_2 \in Z_m$, the *modular distance* between N_1 and N_2, denoted as $D_m(N_1, N_2)$, is given by $W_m(N_1 - N_2)$. Also

$$D_m(N_1, N_2) = W_m(N_1 - N_2) = W_m(N_2 - N_1)$$

In Z_m, it is not true in general that

$$W_m(|N_1 + N_2|_m) \le W_m(N_1) + W_m(N_2) \tag{2.8}$$

The triangular inequality (2.8) is an essential property, as we shall see later, in establishing the correspondence between the "minimum distance" of a code and its error detection and correction capabilities.

EXAMPLE

Let $m = 51$, $N_1 = N_2 = 32 \in Z_{51}$. We have then

$$W_m(32 + 32) = W_m(|64|_{51}) = W_m(13)$$
$$= \min(W(13), W(51 - 13)) = 3$$

Also $W_m(32) = 1$. Therefore

$$W_m(|32 + 32|_m) > W_m(32) + W_m(32)$$

which contradicts (2.8). However, we have the following important theorem.

THEOREM 2.1

When $m = r^n$ (as in radix complement systems) or when $m = r^n - 1$ (as in diminished radix complement systems) the triangular inequality (2.8) holds in Z_m.

Proof

Let N_i' denote N_i or $-\overline{N}_i = N_i - m$, for $i = 1, 2$ correspondingly, whichever has the smallest arithmetic weight. This means that

$$N_1' + N_2' \equiv N_1 + N_2 \,(\text{mod } m), \qquad |N_i| < m$$

and

$$W(N_i') = W_m(N_i) \qquad \text{for} \quad i = 1, 2 \tag{2.9}$$

Let the minimal form expansions of N_1' and N_2' be

$$N_1' = \sum_{i=0}^{n-1} a_i r^i, \qquad |a_i| < r$$

$$N_2' = \sum_{i=0}^{n-1} b_i r^i, \qquad |b_i| < r \tag{2.10}$$

Let S represent the sum of N_1' and N_2' modulo m such that $|S| < m$. In other words,

$$S = \sum_{i=0}^{n-1} c_i r^i \equiv \sum_{i=0}^{n-1} (a_i + b_i) r^i \,(\text{mod } m) \qquad |c_i| < r \tag{2.11}$$

Thus the sum modulo m of N_1' and N_2' can be obtained by the propagation of carries or borrows as required. A carry or borrow from the most significant position will be of magnitude r^n, which equals m (for the radix complement case) or $m + 1$ (for the diminished radix complement case), and therefore can be discarded or propagated as an end-around-carry or end-around-borrow. Consequently, the number of nonzero c_i's in the expression for S can be no greater than the total number of nonzero terms in (2.10). Therefore the number of nonzero terms in S represents the modular weight of the sum $N_1' + N_2'$ and hence also equals the modular weight of $|N_1 + N_2|_M$. Therefore the inequality (2.8) holds.   Q.E.D.

Massey and Garcia [6] show that the triangle inequality (2.8) holds for $m = 2^n \pm 1$. By arguments similar to those in the proof of Theorem 2.1, one can show that (2.8) holds also for $m = 2^n - 2^j \pm 2^i$ $(n - 1 > j > i)$, i.e., for two end-around-carries or borrows. Therefore we formally state the following

COROLLARY 2.2

When $m = 2^n - 2^j \pm 2^i$ for $n - 1 > j > i$, the triangle inequality (2.8) holds in Z_m.

A note on distance and metric

It is important to note that the arithmetic distance and the modular distance between integers are analogous to the well-known *Hamming distance* between vectors or codewords in algebraic linear codes (see, for instance [2, Chapter 1]). In the algebraic linear codes, the minimum Hamming distance of a code relates to its error correcting properties. In Chapter 4, we derive a similar relationship of *minimum arithmetic distance* of an *AN code* to its error control properties.

In the use and understanding of the word *distance* some caution is needed. Distance normally implies a real quantity satisfying the properties of a *metric* which are as follows:

$D(x, y) \geq 0$ equality holds iff $x = y$ (positive definite)

$D(x, y) = D(y, x)$ (symmetry)

$D(x, y) + D(y, z) \geq D(x, z)$ (triangular inequality)

While Hamming and arithmetic distances are invariably metrics, the modular distance is a metric only when the modulus m satisfies the carry properties as required by Theorem 2.1 or Corollory 2.2. The carry properties for a given m depend very much on the radix r of the number system. Thus in the concept of modular distance, the modulus m, the radix r, and the number of end-around caries are involved.

For the purpose of this book, we assume that our interest is mainly in such m and r that do not violate the metric properties. That means, the finite rings Z_m we are concerned with are assumed to be metric spaces. (Exceptions to this rule will be appropriately stated in the text.) In this context, we use the term *modular distance*.

Error sets

We consider R, R', and E in (2.3) to be integers from Z_m, and proceed to define the classes of errors in Z_m. We denote the set of all errors in Z_m of modular weight equal to d by $V(m, d)$, and the set of all errors of weight less than or equal to d by $U(m, d)$. The set of all errors of modular weight 1 is called the set of single errors; modular weight 2, the set of double errors, and so on. Also, $U(m, d)$ equals the union of the sets $V(m, 1)$, $V(m, 2)$, ..., $V(m, d)$.

EXAMPLES

$V(16, 1) = U(16, 1) = \{1, 2, 4, 8, 12, 14, 15\}$

$V(31, 1) = U(31, 1) = \{1, 2, 4, 8, 16, 15, 23, 27, 29, 30\}$

$V(31, 2) = \{3, 5, 6, 7, 9, 10, 11, 12, 13, 14, 17, 18, 19, 20, 21, 22,$
$\qquad\qquad 24, 25, 26, 28\}$

$U(31, 2) = \{V(31, 1), V(31, 2)\} =$ all nonzero elements of Z_{31}

2.3 ERROR CONTROL TECHNIQUES

Logic faults in processors are caused by a number of possible events. Component failures and malfunctions are the most common; overheating, electromagnetic radiation, noise interference, and mechanical shocks are some of the other causes. Logic faults generate errors in the results. These errors have been characterized and classified in the preceding section according to their modular weight.

We discuss here briefly some of the techniques that are employed to detect and correct these errors. These are also referred to as *fault-tolerance* techniques [6] or error control techniques. The objective of these techniques is to obtain a processor that is capable of performing accurately despite the occurrence of logic faults, and thereby increase the reliability, availability, or dependability of the system.

Reliable performance of digital processors is attainable by the systematic application of two techniques.

The first technique involves selection of long-life, highly reliable components; provision of liberal margins between component ratings and the actual operating conditions; and application of proven methods for the interconnection and packaging of these component parts.

The second technique is termed *protective redundancy*, and as the name suggests, it requires the use of redundant equipment (hardware and/or software). The redundant equipment is so designed to detect and/or correct, "bypass," or "mask" the effects of the logic faults. The protective redundancy techniques may be further divided into two classes: *massive redundancy* and *selective redundancy*.

In the massive redundancy approach, the logic faults of a module, such as a component, network, or processor, are masked by permanently connected and parallel operating (fault-free) replicas of the faulty module. As an example, a triplicated processor with the outputs applied to majority vote-takers would mask any output errors due to a faulty processor (see Figure 2.3). This approach is commonly known as triple modular redundancy (TMR) [7, 8], and is discussed in the next section. Other massive redundancy techniques are *quadded logic* as discussed by Tryon [9], *recursive nets* [10], and *adoptive logic elements* [11], among others.

In the category of selective redundancy we include all those that require only a limited (or fractional) increase in hardware or software. Error detection followed by fault diagnosis, and error detection and correction through coding come under this category. Errors are detected by the provision of error-detecting codes, or special monitoring circuits, or by a periodic check through *fault-recognition* or *diagnostic* programs. Use of hardware circuits for error detection has the advantage of instantaneous or *concurrent diagnosis* which may prevent system deterioration or propagation of errors by calling for an appropriate corrective action. *Periodic diagnosis* requires an elaborate corrective action, a "rollback" of the program, and interruption of the system operation for considerable periods of time. When faults are detected by the fault detection circuits, or programs, a corrective action which eliminates the errors is followed. This corrective action will be in the nature of one of the following:

1. error correction by use of error-correcting codes and associated special-purpose error-decoding hardware and/or software;
2. diagnosis to determine the faulty element or module and its replacement by a standby spare [12, 13];
3. reorganization or reconfiguration of the system to enable the necessary computation of functions, if not in a normal mode, in a degraded mode (with loss of precision or speed). This method of operation is often called *graceful degradation*.

There are some good books [7, 14, 17] and papers available on the many different fault-tolerance techniques. This book is intended as a summary of the recent advances in the theory and application of error control codes for arithmetic processors. The arithmetic codes described here are most appropriate for arithmetic operations such as ADD, SHIFT, COMPLEMENT, MULTIPLY, etc. They can also be useful in the control of errors in transmission or in communication channels. The arithmetic codes have a number of properties that parallel and are analogous to those of such communication codes as Hamming codes [15] and cyclic codes [2], but they have many distinct properties of their own, and differ significantly in their algebraic structures.

Triple modular redundancy (TMR)

A TMR system is constructed from a nonredundant system S_0, replacing each logic block in S_0 by three identical blocks and combining their outputs in one or more majority vote-takers. (See Figure 2.5.)

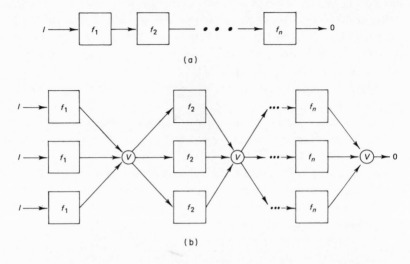

Figure 2.5   (a) Nonredundant system S_0. (b) TMR version of (a).

Figure 2.5a represents a series-connected system with n functional units $f_1, f_2, \ldots, f_n$. Let us assume that the reliability of the unit f_i (for $i = 1, 2, \ldots, n$), which is the probability that the unit f_i is functioning at a given time t, is P_i. The nonredundant system reliability R_0 is given by

$$R_0 = \prod_{i=1}^{n} P_i \tag{2.12}$$

In the redundant system, if we assume the voters to be perfect (i.e., their reliability to be 1), then the reliability of the ith voted triplet is

$$P_i^3 + 3P_i^2(1 - P_i) = 3P_i^2 - 2P_i^3$$

The reliability of the redundant system of Figure 2.5b is

$$R = \prod_{i=1}^{n} (3P_i^2 - 2P_i^3) \tag{2.13}$$

If we assume each functional unit to be of the same reliability, i.e., $P_i = P$ for all i, then

$$R_0 = P^n, \qquad R = (3P^2 - 2P^3)^n$$

If the reliability of the redundant system is to be better than that of the nonredundant system, then

$$R > R_0$$

or

$$(3P^2 - 2P^3)^n > P^n$$

or

$$3P^2 - 2P^3 > P$$

Since P is positive, it means that

$$3P - 2P^2 > 1$$

or

$$(2P^2 - 3P + 1) < 0, \quad \text{i.e.,} \quad (2P - 1)(P - 1) < 0$$

or

$$P > 0.5$$

As long as the reliability of each functional unit is better than 0.5, we could expect improvement in reliability by TMR. This statement is made, of course, under the unrealistic assumption that the voters are perfect. With imperfect voters, where each voter V has a reliability q, system reliability R^* is given by

$$R^* = q^n(3P^2 - 2P^3)^n \qquad (2.14)†$$

Reliability improvement results if

$$q^n(3P^2 - 2P^3)^n > P^n$$

or

$$q(3P^2 - 2P^3) > P$$

or

$$q > P\left(\frac{1}{3P^2 - 2P^3}\right), \quad \text{i.e.,} \quad q > \frac{1}{3P - 2P^2}$$

---

† This formula as well as others in this section is only approximate and does not take into account the cancellation of errors due to failures in successive units. Therefore the actual reliability is slightly better than that given by the formula.

Since $3P^2 - 2P^3$ represents the reliability of a triplet, we have $0 < (3P^2 - 2P^3) < 1$. This leads us to conclude that q must be sufficiently greater than P to obtain reliability improvement. The voters that are imperfect contribute to the unreliability of the system by a factor q^n. If this factor is to be improved, a TMR version with triplicated voters is adopted. (See Figure 2.5c.)

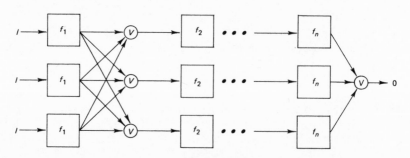

Figure 2.5c

The reliability R^{**} of the TMR system of Figure 2.5c can be obtained as

$$R^{**} = (3P^2 - 2P^3)(3P^2q^2 - 2P^3q^3)^{n-1}q \qquad (2.15)$$

Equations (2.13)–(2.15) can be compared to obtain the conditions as to how P and q must be related to obtain the necessary improvements in reliability. These have been worked out in great detail with charts and figures in the literature [11, 16].

Duplication and switching

Error detection is often a first step in error control coding. Duplication of the essential functional units and matching (or comparing) outputs provide not only the needed error detection, but could also provide the means of establishing an operational system in the presence of logic faults. In case of a mismatch of the outputs from the duplicated units, a corrective action is initiated. The corrective action may be in the nature of a program interrupt, followed by a "system recovery" or a "reconfiguration process" and a diagnosis of the faulty unit. A reconfiguration process involves the location of the faulty unit or network and its replacement by means of "switching" with a standby spare. If a standby spare is not provided, then the reconfigured mode of operation will not provide for matching of outputs, and therefore will be unprotected from future errors. Such a mode of operation is termed *graceful degradation*. Graceful degradation sometimes means that the operational mode is now somewhat degraded in terms of speed, accuracy, or precision.

In summary, the corrective procedures that follow the error detection and eliminate the effects of the fault may be grouped as follows:

1. correction by the replacement of the faulty unit or network by a standby spare, or an active spare [13];
2. system recovery through reconfiguration and switching. The recovered system may operate in a degraded mode until such time as the required maintenance on the failed unit is carried out and it is returned to the system;
3. correction by the application of error-correcting codes such as Hamming [15] or residue [2, 5] codes.

Error-correcting codes have been found to be useful in the detection and correction of errors. These codes can be divided into two major categories:

1. parity-based codes or communication codes;
2. residue codes or arithmetic codes.

The communication codes [2, 15] are by far better known and well developed relative to the arithmetic codes. They are well suited for control of errors during transmission of data over a communication channel or for transmission of data from one section of a computer to another, but not for arithmetic operations. The arithmetic codes, on the other hand, are ideally suited for arithmetic operations and arithmetic processors and can also be used to control errors during data transmission. In the next chapter we introduce fundamentals of arithmetic codes.

PROBLEMS

1. Obtain the arithmetic weight of the following numbers in radix (a) $r = 2$, (b) $r = 3$, (c) $r = 10$ systems:

   (i) 29,   (ii) 199,   (iii) 100.

2. Given $m = 61$, obtain the modular weight ($r = 2$) of numbers 29, 13, and 53.

3. Obtain the error sets $V(61, 1)$, $V(63, 2)$, and $U(63, 2)$. (Use $r = 2$.)

4. Show that the triangular inequality given by (2.8) holds for $m = 2^n - 2^k - 2^j$ for positive integers j, k, n such that $j < k < n - 1$.

5. Given $P_i = P = 0.8$, for $i = 1, 2, \ldots, 5$, $n = 5$, and voter reliability $q = 0.9$, calculate the system reliability for the three cases given by Figures 2.5a–c.

6. Given $n = 2$ for Figures 2.5b and c, derive a condition that makes $R^{**} < R^*$.

7. A metric function d is a real-valued function satisfying

   (i) $d(x, y) \geq 0$, $d(x, y) = 0$ iff $x = y$ (positive definite),
   (ii) $d(x, y) = d(y, x)$ (symmetric),
   (iii) $d(x, y) \leq d(x, z) + d(z, y)$ (triangle inequality).

   Show that the arithmetic distance (D) and the modular distance (D_m) are metrics. Assume $m = 2^n - 2^j \pm 1$ ($n - 1 > j > 1$).

REFERENCES

1. A. Avizienis, Arithmetic Error Codes: Cost and Effectiveness Studies for Application in Digital Systems. *IEEE Trans. Comput.* **C-20**, 1322–1330 (November 1971).

2. W. W. Peterson and E. J. Weldon Jr., "Error-Correcting Codes," 2d Ed., MIT Press, Cambridge, Massachusetts, 1972.

3. G. H. Reitwiesner, Binary Arithmetic, *Advances in Comput.* 1, 232–308 (1960).

4. O. N. Garcia, Error Codes for Arithmetic and Logical Operations. Ph.D. Thesis, Dept. of Elec. Engrg., Univ. of Maryland, College Park, 1969.

5. T. R. N. Rao and O. N. Garcia, Cyclic and Multi-residue Codes for Arithmetic Operations, *IEEE Trans. Information Theory* **IT-17**, 85–91 (January 1971).

6. J. L. Massey and O. N. Garcia, Error correcting codes in computer arithmetic, Chapter 5, in "Advances in Information System Sciences" (J. L. Tow, ed.), Vol. 4, pp. 273–326. Plenum Press, New York, 1971.

7. W. H. Pierce, "Failure-Tolerant Computer Design." Academic Press, New York, 1965.

8. J. Von Neumann, "Probabilistic Logics and the Synthesis of Reliable Organisms from Unreliable Components" (Automata Studies, Ann. Math. Studies, No. 34). Princeton Univ. Press, Princeton, New Jersey, 1956.

9. J. G. Tryon, Quadded Logic, *in* "Redundancy Techniques for Computing Systems" (R. H. Wilcox and W. C. Mann, eds.). Spartan Books, Washington, D.C., 1962.

10. S. Levy, Reliability of Recursive Triangular Switching Networks Built of Rectifier Gate, *in* "Redundancy Techniques for Computing Systems" (R. H. Wilcox and W. C. Mann, eds.). Spartan Books, Washington, D.C., 1962.

11. W. H. Pierce, Adaptive Vote-taxers Improve the Use of Redundancy, *in* "Redundancy Techniques for Computing Systems" (R. H. Wilcox and W. C. Mann, eds.). Spartan Books, Washington, D.C., 1962.

12. A. Avizienis, Design of Fault-Tolerant Computer, *Proc. Fall Joint Comput. Conf., 1967*, pp. 733–743.

13. R. W. Downing, *et al.*, No. 1 ESS Maintenance Plan, *Bell System Tech. J.* pp. 1961–2019 (September 1964).

14. "Redundancy Techniques for Computing Systems" (R. H. Wilcox and W. C. Mann, eds.). Spartan Books, Washington, D.C., 1962.

15. R. W. Hamming, Error Detecting and Error Correcting Codes, *Bell System Tech. J.* **29**, 147–160 (April 1960).

16. D. K. Rubin, The Approximate Reliability of Triply Redundant Majority-voted Systems, *Proc. of First Annu. IEEE Comput. Conf., Chicago, Illinois, September, 1967*.

17. F. F. Sellers, Jr., M. Y. Hsiao, and L. W. Bearnson, "Error Detecting Logic for Digital Computers." McGraw-Hill, New York, 1968.

3 ARITHMETIC CODES, THEIR CLASSES AND FUNDAMENTALS

An arithmetic error code for us is a redundant representation of numbers having the property that certain errors can be detected and/or corrected in arithmetic operations using these numbers. The representation is redundant in that the number of digits used for representing a number in a coded form may be larger than the minimum number of digits required if no error control is desired. The fundamental arithmetic operation is addition. Therefore, any useful arithmetic code must at least check addition. Preferably, all other elementary operations, such as shift and cycle, should be checked as well.

3.1 CODE CLASSES

To represent the set of integers Z_m in the radix r system, the number of digits required, k, is the smallest integer greater than or equal to $\log_r m$. This is denoted by

$$k = \lceil \log_r m \rceil \dagger \qquad (3.1)$$

† $\lceil x \rceil$ denotes the smallest integer greater than or equal to x.

Instead of using k digits as minimally required to represent Z_m, a redundant code may use n digits for some $n > k$. This may be in the nature of adding an extra $n - k$ digits as checks to the nonredundant form of k digits; or it may be to denote each number $x \in Z_m$ by a product AX for some constant integer A. Since these codes are to check arithmetic operations, it is important to define how these operations are carried out on redundant forms. Depending on how a number $N \in Z_m$ is represented as an n-tuple or how arithmetic is performed on the codewords, the codes are classified as separate or nonseparate, and as systematic or nonsystematic. All in all, there are three classes of codes, each important in its own right. These are (1) *AN codes*, which are nonseparate and nonsystematic; (2) *separate codes*, which are always systematic; and (3) *systematic but nonseparate codes*.

Before defining these code classes let us look at the examples of binary codes given in Table 3.1. For the integers N whose decimal and binary forms appear in columns 1 and 2, respectively, three different binary code representations are shown in columns 3–5. These three codes can be called equivalent in the sense that their error control properties are the same. They differ, of course, in their representation and practical implementation, as we shall observe later. In column 3 of Table 3.1 we have a $5N$ code which is an example of the class of codes

Table 3.1 Examples of binary codes

Decimal N	Binary N	Binary arithmetic codes representing N		
		$5N$ code	$[N, |N|_5]$ code	$|25N|_{40}$ code
0	0 0 0	0 0 0 0 0 0	(000, 000)	0 0 0 0 0 0
1	0 0 1	0 0 0 1 0 1	(001, 001)	0 1 1 0 0 1
2	0 1 0	0 0 1 0 1 0	(010, 010)	0 0 1 0 1 0
3	0 1 1	0 0 1 1 1 1	(011, 011)	1 0 0 0 1 1
4	1 0 0	0 1 0 1 0 0	(100, 100)	0 1 0 1 0 0
5	1 0 1	0 1 1 0 0 1	(101, 000)	0 0 0 1 0 1
6	1 1 0	0 1 1 1 1 0	(110, 001)	0 1 1 1 1 0
7	1 1 1	1 0 0 0 1 1	(111, 010)	0 0 1 1 1 1

popularly called AN codes. (Here $A = 5$.) Each N in the range $0 \leq N < m = 8$, requiring only three binary digits (or bits) in a nonredundant representation, will now require six bits in the $5N$ code. Over the six-digit positions of the $5N$ code, no group of three fixed positions exists which can be identified explicitly as the integer N. On the other hand, in column 4 is a code labeled $[N, |N|_5]$, where the integer N and the check $|N|_5$ together still require only six bits but are separated by a comma, so that the information part is clearly identified.

In column 5 of the table is yet another code, labeled $|25N|_{40}$. In this code, the codeword (for any integer N) is the product $25 \times N$ in the ring of integers modulo 40. We later discuss the details as to how these numbers, 25 and 40, have been arrived at, but for the present it is easy to observe that each codeword is a distinct multiple of 5 and is less than 40 in magnitude. It is easy to verify that this code is a permutation of the $5N$ code (of column 3). However, there is a striking difference between the $5N$ code and this code. The integer N can be identified from the rightmost three bits in the $|25N|_{40}$ code. Now keeping in mind that this is a form of $5N$ code, we can obtain the numbers 25 and 40 as follows: 25 equals $bm + 1$, for some $b \geq 1$, and $40 = 5m$. In fact, we will establish later that for any AN code, $0 \leq N < m = r^k$, we can obtain an equivalent code of the form $|gAN|_M$, where $M = Am$ and $gA = bm + 1$ for a suitable integer b $(1 \leq b < A)$. This, however, is conditional on A and m being relatively prime. A rigorous discussion with proofs is provided in Chapter 7. We must take note, however, that each codeword here is just one integer and has no separate parts, and hence is not a separate code. Further, in the example of the $|25N|_{40}$ code, two codewords are to be added as in the ring of integers modulo 40. The structure then preserves closure under addition and also enables error detection just as in the $5N$ code or the $[N, |N|_5]$ code. Using these examples as a basis, we can formally define the code categories as follows.†

---

† The systematicity defined here is quite standard and is same as in communication codes. (See Peterson [4, Chapter 3].) The definition of separateness is due to Peterson [4, Chapter 15] and Garner [5].

DEFINITION 3.1

An arithmetic code, which has each codeword represented by, say, n digits, is systematic if there exists a set of k digits $(k < n)$ of the codeword representing the information and the remaining $n - k$ digits representing the check(s).

A systematic code may treat the two parts, the information digits and the check digits, separately for the purpose of addition, thereby defining two or more independent addition structures, one for the information and the others for the checks; or it may treat each codeword as a single operand (or number) and define uniform addition rules for all the n digits except perhaps for some end-around-carries. A systematic code of the former type is called separate, and the latter nonseparate. A similar division into separate and nonseparate classes can be made for all codes. Since a code that has separate independent addition structures for the information and checks must also be systematic, we have only the three classes of codes as mentioned previously. By these definitions, the $[N, |N|_5]$ and $|25N|_{40}$ codes are systematic, whereas the $5N$ code is nonsystematic. Only the code $[N, |N|_5]$ of Table 3.1 is of the separate type, the other two being nonseparate.

3.2   *AN* CODES AND SINGLE-ERROR DETECTION

We introduce some fundamentals of a class of arithmetic codes first studied by Diamond [1] and Brown [2] as *AN codes*, and later by Chien [3], Peterson [4], Garner [5], and others [6, 7]. These codes have also been referred to as *linear residue* codes [3], *product* codes [6], and *residue-class* codes [7].

In an *AN* code, a given integer N is represented by the product $A \cdot N$ for some suitable constant integer A. A is commonly called the *generator* (and sometimes *check base*) of the code. Naturally, the range for N may be limited to $0 \le N < m$ (or $-m/2 \le N < m/2$), where m

is the *cardinality* or *range* of the code information. Depending on the radix r used for representing the products, an *AN* code is called, for example, binary ($r = 2$), ternary ($r = 3$), or decimal ($r = 10$).

The number of digits required to represent a codeword is said to be the *code length*. That is, each of the products $0, A, 2A, \ldots, A(m-1)$ is an n-digit representation in radix r. Since k represents the number of digits required for information N, the number of redundant digits in the code is given by $n - k$. Sometimes the redundancy of the code is expressed by the value $\log_r A$, which differs from $n - k$ by less than 1.

Consider the addition of two integers N_1 and N_2. The sum of their coded forms, $AN_1 + AN_2$, is equal to $A(N_1 + N_2)$, the coded form of their sum. The result of the addition is a codeword, provided, of course, that $N_1 + N_2 < m.$† Hence, the sum can be checked for errors. An error E in the addition will result in an actual sum

$$S = A(N_1 + N_2) + E = AN_3 + E \tag{3.2}$$

The check is to verify whether the result obtained is a codeword, as is to be expected. This verification can be accomplished by division of S by A and by checking whether the remainder is zero. This, in mathematical terms, is to obtain the least nonnegative residue of S modulo A, denoted as $|S|_A$. Therefore,

$$|S|_A = |AN_3 + E|_A = |AN_3|_A + |E|_A = |E|_A$$

If there is no error, i.e., $E = 0$, then $|E|_A = 0$. Also if E is an exact multiple of A, $|E|_A = 0$; hence, the check fails, and we call such errors undetectable. There is really no way to tell whether there is no error at all or the error is undetectable. However, the probability of occurrence of an undetectable error is usually very small if A is chosen properly. If any error E in a parallel binary adder is of the type $\pm 2^j$ for some j, then $A = 3$ will be able to detect any such error. Also, since

---

† The code will be closed under addition for all cases (i.e., including the case $N_1 + N_2 \geq m$) if the addition of the code words AN_1 and AN_2 is given by $|AN_1 + AN_2|_M$ for $M = A \cdot m$, as is shown later.

$|\pm 2^j|_3 \neq 0$ for all values of j, single errors can be detected no matter what the length of the code is. This result is analogous to the case of using a parity bit for single-error detection in block codes for communication systems.

Generalization of the result above to a nonbinary code such as a ternary or a decimal code is straightforward. A single error E in a radix r representation is of the type $\pm b r^j$, where $1 \leq b < r$. Since $|E|_A = |\pm b r^j|_A \neq 0$ for single-error detection, A can be any integer which contains a factor that is greater than and relatively prime to r. Obviously the choice of $A = r + 1$ will always work. $4N$ ternary and $11N$ decimal codes are natural examples.

A class of codes useful for error detection is obtained by selecting $A = r^c - 1$ for some integer $c > 1$. For the binary case, these codes turn out to be $A = 3, 7, 15$, or 31, etc. Their proven usefulness is due to the property that the residue modulo A of a binary number N can be obtained without actual division. As an example, let $N = 10110010101$. $|N|_3$ is obtained by repeated addition of the two-bit bytes of N with an end-around-carry as follows

$$N = 1|01|10|01|01|01$$

01	10
01	01
10	11
01	01
11	100
10	1
101	01
1	
10	

$|N|_3 = |010110010101|_3 = |1429|_3 = 1$, as can be verified. For obtaining the residues modulo $r^c - 1$ of an n-digit radix r integer a similar procedure can be derived as follows [8].

Let $x = x_{n-1} \cdots x_0$ be an n-digit integer in radix r. Let the n-digit integer be partitioned into l bytes of length c where $lc \geq n > (l - 1)c$. Let these bytes be $B_{l-1} \cdots B_1 B_0$. Then we have

$$x = \sum_{j=0}^{l-1} B_j r^{jc} \qquad \text{where} \quad 0 \leq B_j < r^c$$

$$|x|_A = |x|_{r^c-1} = \left| \sum_{j=0}^{l-1} B_j r^{jc} \right|_{r^c-1}$$

(3.3)

Since $|r^{jc}|_{r^c-1} = 1$ for all $j \geq 0$, we have

$$|x|_A = \left| \sum_{j=0}^{l-1} B_j \right|_{r^c-1}$$

(3.4)

The above provides the basis for obtaining the residue generator schematics given in Section 3.5.

3.3 CHECKING AN ADDER BY SEPARATE CODES

Error detection of a binary adder by an AN code has been discussed in the previous section. A separate code of the type $[N, |N|_3]$ can also be conveniently used for error detection. Before considering such a code for detection of errors in addition and other elementary operations, we establish here that a separate code, meaning a code whose information and checks are separately processed, must be a residue code. This result is obtained by Peterson by considering a separate adder and checker as follows (see Figure 3.1) [4, 9].

Let the check corresponding to the information N be denoted by $C(N)$, thus forming a codeword $[N, C(N)]$. Also let us define addition of two codewords to yield

$$[N_1, C(N_1)] + [N_2, C(N_2)] = [N_1 + N_2, C(N_1) * C(N_2)] \qquad (3.5)$$

Note here that addition of codewords involves the two operations $+$ and $*$ on the information and check parts, respectively. Further, if the equation

$$C(N_1) * C(N_2) = C(N_1 + N_2) \qquad (3.6)$$

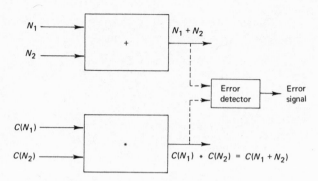

Figure 3.1 Separate adder and checker.

is satisfied, then the sum of any two codewords yields another codeword. That is, the code is closed under addition. The error checking will now constitute simply checking for whether the sum is a codeword. This check is the function of the error detector in Figure 3.1 whose inputs are the outputs of the adder and the checker, and whose output is an error signal. Peterson [9], using N_1 and N_2 from the infinite set of all integers and assuming that the number of check symbols $C(N)$ is fewer than the number of information symbols, which only means that the number of check digits required for $C(N)$ is finite, obtained the following important result.†

THEOREM 3.1

If there are fewer check symbols than integers permissible in the range of N, and if the check symbols $C(N)$ satisfy (3.6), then $C(N)$ must be the residue of N modulo some base b and $*$ is addition modulo b of these check symbols.

By virtue of this result every separate error code that is closed under addition must be a separate code of the form $[N, |N|_b]$. There is no

---

† The proof of the theorem is available either from Peterson [4] or from [9]. A formal extension of this theorem is stated next and proved somewhat along the lines of that given in these references.

restriction placed on the values of N_1 and N_2 and the addition structure in the discussion above. But in practical binary adders N_1 and N_2 are n-bit numbers and can therefore be conveniently treated as elements of Z_m, and their sum as $N_1 + N_2$ modulo m. (Once again we let $m = 2^n$ or $2^n - 1$, depending on whether 2's complement or 1's complement addition is employed.) A generalization of Peterson's result will be given following the next lemma.

LEMMA 3.2

For any N, $||N|_x|_y = |N|_y$ if and only if y divides x.

Proof

Let y divide x. Then $|x|_y = 0$. Since $N \equiv |N|_x$ (mod x), it follows, that $|N|_x = N + kx$ for some k. Therefore

$$||N|_x|_y = |N + kx|_y = ||N|_y + |kx|_y|_y = |N|_y$$

Conversely, if $||N|_x|_y = |N|_y$ for all N, then $|kx|_y = 0$. Since k is arbitrary, $|x|_y = 0$ or y divides x.

THEOREM 3.3

Let N_1 and N_2 in Figure 3.1 be elements of Z_m, and let $\oplus$ denote addition in Z_m. Assume $C(N)$ needs no more digits than $N \in Z_m$ requires for its representation. Then,

$$C(N_1 \oplus N_2) = C(N_1) * C(N_2) \tag{3.7}$$

holds if and only if for some b

(i) $C(N) = |N|_b$,
(ii) $*$ denotes addition modulo b, and   (3.8)
(iii) b divides m.

Proof

For an $[N, C(N)]$ code satisfying (3.8), the code can be written as an $[N, |N|_b]$ code. The sum of two words $[N_1, |N_1|_b], [N_2, |N_2|_b]$ results in

$$[N_1 \oplus N_2, ||N_1|_b + |N_2|_b|_b]$$

Note here that $N_1 \oplus N_2$ in Z_m is the same as $|N_1 + N_2|_m$ in Z.† Hence

$$C(N_1) * C(N_2) = ||N_1|_b + |N_2|_b|_b = |N_1 + N_2|_b$$

Also since b divides m, we obtain

$$C(N_1 \oplus N_2) = C(|N_1 + N_2|_m) = ||N_1 + N_2|_m|_b = |N_1 + N_2|_b$$

by Lemma 3.2 and that establishes (3.7).

This proves that a separate residue code whose check base b divides the range m of the adder is closed under addition. It remains to show that if a separate $[N, C(N)]$ code is closed under addition, then it is a separate residue code. To accomplish this we must show that (3.7) implies (3.8).

Let S denote the subset of Z_m such that for any $N \in S$, $C(N)$ is the same as $C(0)$. Obviously $0 \in S$, and for any $x, y \in S$, $C(x) = C(y) = C(0)$. Also $C(x) = C(x \oplus 0) = C(x) * C(0)$. For any $x, y \in S$, by Equation (3.7)

$$C(x \oplus y) = C(x) * C(y) = C(0) * C(0) = C(0 \oplus 0) = C(0)$$

and therefore $x \oplus y \in S$. Thus S is closed under $\oplus$. Also for $n \in Z_m$ and $x \in S$, the product $nx = x \oplus x \oplus \cdots \oplus x$, a total sum of n times, is in Z_m.

$$C(nx) = C(x \oplus (n-1)x) = C(x) * C((n-1)x)$$
$$= C(x) * C(x) * C(x) * \cdots * C(x)$$

a total of n times. Since $x \in S$, $C(x) = C(0)$ and therefore $C(nx) = C(0) * C(0) * \cdots * C(0) = C(0)$, and hence $nx \in S$, which makes the subset S an ideal in Z_m. Also Z_m is a principal ideal ring in the sense

† Z is the set of all integers.

that every ideal in Z_m is a principal ideal. In other words, S has a unique generator, say, b, and all the elements of S are multiples of b. From the algebraic theory of rings and ideals, we also know that b is the smallest positive integer in S, and that b also divides m.

If $b = 1$, the ideal S is the same as the ring Z_m, and the check for any $N \in Z_m$ is $C(N) = C(0)$, making the code useless. Next we observe that $C(N_1) = C(N_2)$ if and only if N_1 and N_2 are in the same residue class of S in Z_m as follows. Assume that N_1 and N_2 are in the same residue class of S. Then $N_1 - N_2$ is in S and $C(N_1 - N_2) = C(0)$. (Note here that $-N$ in Z_m is the additive inverse of N, and therefore denoting $-N$ by $\overline{N}$, we can write $N_1 - N_2 = N_1 \oplus \overline{N}_2$.)

$$C(N_1) = C(N_2 \oplus N_1 - N_2) = C(N_2) * C(N_1 - N_2)$$
$$= C(N_2) * C(0) = C(N_2)$$

Alternatively, if $C(N_1) = C(N_2)$, then

$$C(N_1 - N_2) = C(N_1 \oplus \overline{N}_2) = C(N_1) * C(\overline{N}_2)$$
$$= C(N_2) * C(\overline{N}_2) = C(N_2 \oplus \overline{N}_2) = C(0)$$

and therefore $N_1 - N_2$ is in S, and N_1 and N_2 are in the same residue class of S. This also means that N_1 and N_2 have the same check symbol if and only if the two differ by an element of S, meaning $N_1 \equiv N_2$ (mod b), where b is the generator of S, as stated previously.

If b is a proper divisor of m, then S has more than one element and so also do the residue classes. For convenience, we may denote the check symbol for the elements of S by 0, which means that $C(0) = 0$. Also for every j, $0 < j < b$, let $C(j) = j$ denote the check for the entire residue class containing j. Then $i * j = C(i) * C(j) = C(i \oplus j) \equiv i + j$ modulo b; that is, the operation $*$ is the same as addition modulo b.

Q.E.D.

EXAMPLES

If N_1 and N_2 are elements of Z_m, for $m = 2^{10} - 1$ (as in 1's complement 10-bit system), then the $[N, |N|_b]$ code is closed under addition for any b that is a divisor of m. Naturally b can be any one of 3, 11, 31,

33, 93, 341, 1023. If $b = 1023$, the checker in Figure 3.1 becomes a mere duplication of the adder, and the error detector is a match (or comparison) of the two input numbers.

On the other hand, if N_1, $N_2 \in Z_m$, $m = 2^{10}$ (for the 2's complement 10-bit system), any $[N, |N|_b]$ code that is closed under addition must have the check base b as a divisor of 2^{10}. That means b is a power of 2, and $|N|_b$ is merely a duplication of a few rightmost bits of N. It is obvious that the code cannot perform any check on the left (most significant) bits, unless $b = m$, and hence it is not very efficient.

Error detector

In the previous section, we established that the check base b of a separate residue code must divide m, the range of the adder, if the code is to be closed under addition. Further, if the code is closed under addition, the outputs of the adder and checker form a codeword, and thereby the function of the error detector becomes one of just checking that relation between R and Q (see Figure 3.2). Therefore, the error detector may consist of a residue generator and a match circuit.

Consider for a moment that b does not divide m. (This is more likely when $m = 2^n$.) The code is not closed under addition for all values of N_1 and N_2, and therefore the outputs R and Q do not form a codeword all the time. However, if some form of corrections can be applied to the

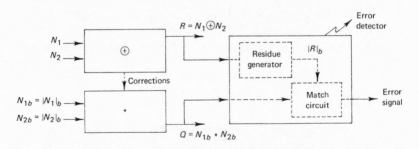

Figure 3.2 Error detection schematic for addition.

checker as inputs from the adder (as indicated by the dashed line in Figure 3.2), then the outputs can still be made to form a codeword, and an error detector such as the one shown in Figure 3.2 can be utilized. This discussion only points out that the checker is simpler if the code is closed under addition than if it is not.

It is important to choose b in such a way that the single errors (i.e., the most likely errors) in the adder and the checker are detected. We operate under the assumption that (1) at most one of the adder and checker units can be erroneous, and (2) the error detector is a perfect device at all times. For clarity let us consider the two separate cases as follows.

1. Error in the adder

We denote the output of the adder by $R' = R \oplus E = N_1 \oplus N_2 \oplus E$, where E represents the error (number). E can be an element of the set $U(m, d)$, for some d, as defined in Section 2.2. In the error detector, the residue of R' modulo b given by

$$||N_1 \oplus N_2|_b + E|_b$$

is matched against $Q = |N_1 \oplus N_2|_b$. An error signal is generated whenever $|E|_b \neq 0$.

2. Error in the checker

Let $Q' = |Q + e|_b = ||N_1 \oplus N_2|_b + e|_b$ denote the (possibly) erroneous output of the checker, where e is the error. Q' matches with $|R|_b$, the error-free output of the adder iff $|e|_b = 0$. In other words, the error e is detectable unless it is a multiple of b. Since the checker generates numbers modulo b, i.e., numbers smaller than b, all errors in the checker are detectable. *Any mismatch detected by the match circuit generates an error signal. But there is no way to find out which of the two units, the adder or the checker, is erroneous.*

EXAMPLES   (Binary error-detecting codes)

Consider the adder in Figure 3.2 to be a 1's complement nine-bit adder. Then N_1, $N_2 \in Z_m$ where $m = 2^9 - 1 = 511$. *The check base b is to be chosen such that the code* $(N, |N|_b)$ *is closed under addition and is able to detect all single errors.* We consider a single error E to be an element of the set $V(m, 1)$. Clearly E is of the form 2^j $(j = 0, 1, \ldots, 8)$ or its complement with respect to 511. The check base b must be such that $|m|_b = 0$ and $|E|_b \neq 0$ for all $E \in V(m, 1)$. The choices for b then are 7, 73, and 511. (Note $511 = 7 \times 73$.)

As another example, we choose $m = 511$ as before, but let E be an element of $U(511, 2)$. That is, the code is to detect all single and double errors. Since $7 \in U(m, 2)$ but $73, 511 \notin U(m, 2)$, 73 and 511 are likely choices. If $b = 511$, the checker is a duplication of the adder and detects all errors. By an examination, 73 and all its multiples in Z_m have a modular weight greater than or equal to 3, and therefore for any $E \in U(511, 2)$, $|E|_{73} \neq 0$. Thus 73 is the only other check base that detects all double (or fewer) errors and is closed under the addition operation.

3.4 CHECKING OTHER ELEMENTARY OPERATIONS

The elementary operations of an arithmetic processor were characterized in Section 1.3, and expressions for the result $R = \Phi(N_1, N_2)$ were derived and tabulated for the two cases of 2's complement and 1's complement (Table 1.4).† We extend here the error-checking schematic of the adder (Figure 3.2) to all other elementary operations [10]. In order to simplify the checking logic, we assume that the check base b is a divisor of m, the range of the numbers in the adder. This means that we use 1's complement addition, and therefore that the range is $m = 2^n - 1$. Further, we assume that b, a divisor of m, is also of the form $2^c - 1$ for some c, and thereby the generation of residues modulo b can be obtained without actual division. That is, as stated

---

† In order to avoid ambiguity between $A = \delta_I(A)$ and the generator A, the operands A and B in the model for an arithmetic processor have been changed here to N_1 and N_2, respectively.

earlier, the generation of residues modulo b can be obtained by grouping into bytes of length c and adding these bytes modulo b with an end-around-carry. Further, we can state the following.

LEMMA 3.4

For positive integers $r \geq 2$ and $c \leq n$, $r^c - 1$ divides $r^n - 1$ iff c divides n.

Proof

Let $b = r^c - 1$. Then $r^c \equiv 1 \pmod{b}$. If c divides n, then $n = ck$ for some k. Also $(r^c)^k \equiv 1^k \pmod{b}$ or $r^{ck} = r^n \equiv 1 \pmod{b}$. Hence $r^n - 1 \equiv 0 \bmod (r^c - 1)$, which means that $r^c - 1$ divides $r^n - 1$. On the other hand, if c does not divide n, then $n = ck + d$ for some k and $0 < d < c$. Then $|r^n|_b = |r^{ck+d}|_b = |r^{ck} \cdot r^d|_b$. Since $|r^{ck}|_b = |(r^c)^k|_b = 1$, we have $|r^n|_b = |r^d|_b$. Further, $r^d < b$ and $|r^d|_b \neq 1$. $|r^n - 1|_b = r^d - 1 \neq 0$. Hence b does not divide $r^n - 1$, and the lemma is proved.

From Lemma 3.4, we have that if the check base $b = 2^c - 1$ is to be a divisor of $2^n - 1 = m$, then c must be a divisor of n. This, in other words, states that the registers (and adder) in the checker are of c bits (or c stages) while the operand registers in the processor are of n bits, such that c divides n. Structurally, the processor and the checker maintain strong similarities.

The error-checking procedure is evident from Figure 3.3. The arithmetic processor generates the result $R = \Phi(N_1, N_2)$, while the checker generates $Q = \Phi_b(N_{1b}, N_{2b})$ where $N_{1b} = |N_1|_b$ and $N_{2b} = |N_2|_b$. Together R and Q form a codeword. In other words, for each elementary operation Φ_i $(i = 1, 2, \ldots, 8)$, a corresponding parallel operation Φ_{bi} is required for the checker to enable the error-checking scheme to work. Since the functions denoted by Φ_i have been characterized before (Table 1.4), our problem here will be one of deriving the appropriate formulas for Φ_{bi}. Although the function Φ_{bi} is written as a function of N_{1b} and N_{2b}, some corrections will be required in the form of A_0 and A_{n-1} from the arithmetic processor to the checker, as can be seen from Figure 3.3 and Table 3.4. Addition of the residues N_{1b} and

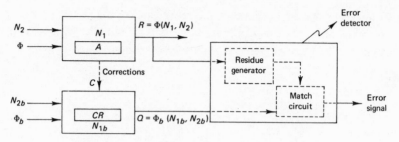

Figure 3.3   Error-checking schematic for operation Φ.

N_{2b} modulo b can be obtained by a conventional c-bit binary adder with an end-around-carry (note that $b = 2^c - 1$). For subtraction of a residue modulo b, the 1's complement of N_{2b} is added to N_{1b}. The operations Φ_{b3}, Φ_{b4}, and Φ_{b7} require some corrections. Multiplication (or division) by 2 of a residue modulo b can be accomplished by left (or right) cyclic shift of the c-bit check register.

Table 3.4[a]

Operation Φ_i	Result $R = \Phi_i(N_1, N_2)$ $m = 2^n - 1$	Operation Φ_{bi} $(b = 2^c - 1)$	Check $Q = \Phi_{bi}(N_{1b}, N_{2b})$
Φ_1 (ADD)	$\lvert N_1 + N_2 \rvert_m$	Φ_{b1}	$\lvert N_{1b} + N_{2b} \rvert_b$
Φ_2 (SUB)	$\lvert N_1 - N_2 \rvert_m$	Φ_{b2}	$\lvert N_{1b} - N_{2b} \rvert_b$
Φ_3 (LSHR)	$\frac{1}{2}(N_1 - A_0(t))$	Φ_{b3}	$\left\lvert \dfrac{N_{1b} - A_0(t)}{2} \right\rvert_b$
Φ_4 (ASHR)	$\frac{1}{2}(N_1 - A_0(t) + A_{n-1}(t)2^n)$	Φ_{b4}	$\left\lvert \dfrac{N_{1b} - A_0(t) + A_{n-1}(t)}{2} \right\rvert_b$
Φ_5 (CYR)	$\lvert \frac{1}{2}N_1 \rvert_m$	Φ_{b5}	$\lvert \frac{1}{2}N_{1b} \rvert_b$
Φ_6 (CYL)	$\lvert 2N_1 \rvert_m$	Φ_{b6}	$\lvert 2N_{1b} \rvert_b$
Φ_7 (SHL)	$\lvert 2N_1 - A_{n-1}(t) \rvert_m$	Φ_{b7}	$\lvert 2N_{1b} - A_{n-1}(t) \rvert_b$
Φ_8 (CLAD)	N_2	Φ_{b8}	N_{2b}

[a] The reader should note here that the N_1 and N_2 used in this table replace the integer operands A and B, respectively, of the earlier model for arithmetic processors. $A_i(t)$ denotes here the initial value (i.e., at time t) of the ith bit of N_1.

It should be pointed out here that if 2's complement arithmetic is used in the AP, a checking procedure can still be employed along the same lines, although the corrections then required from the AP to the checker will increase to some extent.

3.5 RESIDUE GENERATORS

In this section we briefly discuss the techniques required to generate the residue of a binary number N with respect to a check base b. For most of our applications b is of the form $2^c - 1$ for some c. We present an organization of a mod 15 residue generator and also discuss the complexities involved in the design of a residue generator where b is not of the form $2^c - 1$ ($b = 29, 47, 61$, etc.). Such check bases will be of interest in error correction as can be seen in later chapters.

Let N be a 24-bit binary number. We break it into six four-bit bytes. Pairs of bytes (B_0, B_1), (B_2, B_3), and (B_4, B_5) are added using mod 15 adders with end-around-carries until the sum is reduced to a four-bit number. Then the sum will represent the residue of N mod 15. (See Figure 3.4). When b is not of the form $2^c - 1$, the organization

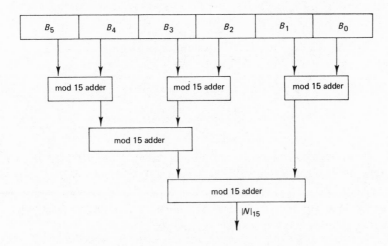

Figure 3.4 Parallel residue generator modulo 15.

is more complicated. Let us consider a specific example of b, say, 29. We give here three different organizations to generate $|N|_{29}$, where N is given by the binary n-tuple $(I_{n-1}, I_{n-2}, \ldots, I_1, I_0)$.

Serial residue generator modulo 29

The shift register **D** of five binary cells can be used to obtain the residue of N mod 29, as shown in Figure 3.5a, where HA and FA are half-adder and full-adder blocks, respectively. Initially the register is to be cleared and the binary number whose residue modulo 29 is required is applied as the input serially with high-ordered digits first. With each input digit advance, a left shift pulse is applied to the register.

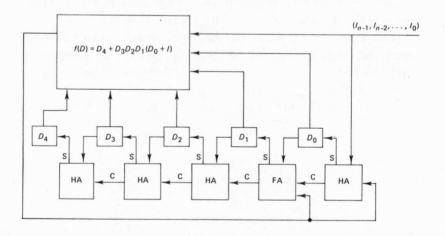

Figure 3.5a Serial residue generator modulo 29.

If the register value at time t is represented by $D(t) = x$ and the input by I, the register value at time $t + 1$, denoted by $D(t + 1)$, is given by $|2x + I|_{29}$. By the time the lowest ordered digit enters the register, the register will have the residue $|N|_{29}$ formed inside, which thus requires a total of n shifts for the generation.

Before we present the other two techniques a small note on $|N|_{29}$ is in order. For a 24-bit number N, divided into five bytes of five bits each, $|N|_{29}$ can be expressed as

$$N = (B_4, B_3, B_2, B_1, B_0)$$
$$|N|_{29} = |32^4 \cdot B_4 + 32^3 B_3 + 32^2 B_2 + 32 \cdot B_1 + B_0|_{29}$$
$$= |3^4 B_4 + 3^3 B_3 + 3^2 B_2 + 3B_1 + B_0|_{29}$$
$$= |3(3(3(3B_4 + B_3) + B_2) + B_1) + B_0|_{29}$$

This forms a basis for the following residue generation schemes.

Serial-by-byte residue generation modulo 29

In Figure 3.5b is shown a serial-by-byte residue generator. Initially the register **CR** is cleared. When the first clock pulse is applied, the contents of B_4 are added to $|3 \times CR|_{29}$ in a mod 29 adder and the sum is transferred to **CR**. At the same time the contents of register **B** are cycled by one byte. At the next clock pulse the contents of B_4 (previously the contents of B_3) are added to $|3 \times CR|_{29}$ in the mod 29 adder, the sum is transferred to **CR**, and the register **B** is left cycled by one byte.

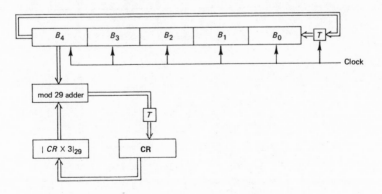

Figure 3.5b Serial-by-byte residue generator mod 29.

By repeating this procedure, the residue $|N|_{29}$ will be formed in **CR** in five clock pulses and the contents of **B** will also be restored to their original value N.

Parallel residue generator modulo 29

A parallel organization of the residue generator is shown in Figure 3.5c. Let the number N whose residue modulo 29 is required be divided into five bytes. The bytes B_4, B_3, B_2, B_1, B_0 are multiplied by their weights 23, 27, 9, 3, 1, respectively. $|B_4 \times 23|_{29}$ and $|B_3 \times 27|_{29}$ are added in a mod 29 adder to obtain the sum S_1. $|B_2 \times 9|_{29}$ and $|B_1 \times 3|_{29}$ are added to obtain S_2. S_1 and S_2 are added in a mod 29 adder to obtain the sum S_3 which is added to B_0 in a mod 29 adder. The sum obtained will be $|N|_{29}$. From these examples, it is clear that the parallel residue generation becomes complex when b is not of the form $2^c - 1$.

Residue generation (or syndrome generation) can be obtained by suitable table look-up procedures. For large numbers, the residue table, i.e., the read-only memory (ROM) size, is prohibitively large and

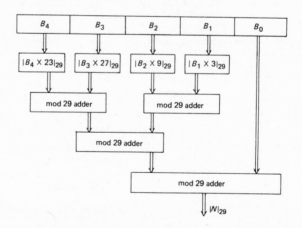

Figure 3.5c   Parallel residue generator modulo 29.

impractical. However, the number whose residue modulo some b is required can be divided into bytes of fixed size, and the residues of these bytes can be obtained by a table look-up using only reasonably sized ROM. Then residues so obtained have to be multiplied by an appropriate constant before they can be added modulo b. Multiplying by a constant is an appropriate fixed permutation of the address decoder wires of the ROM. Also addition of the residues modulo b could also be performed by table look-up techniques. This approach for residue generation has been discussed in detail and its feasibility has been explored by Sitnichenko [11].

Match circuit

The function of a match circuit is to compare two inputs and generate a signal whenever the inputs differ. For our purpose we can utilize an EXCLUSIVE-OR logic block as the match circuit. The residue $|R'|_b = |N_1 + N_2 + E|_b = ||N_1 + N_2|_b + E|_b$ is compared with $Q = |N_{1b} + N_{2b}|_b$ in the match circuit. An error signal is generated whenever $|E|_b \neq 0$. Similarly the errors in the checker are detected.

PROBLEMS

1. For a given AN code, $0 \leq N < m \leq r^k$, an equivalent systematic nonseparate code can be obtained as $|gAN|_M$ code, where $M = A \cdot m$ and $gA = bm + 1$ for some b such that $1 \leq b < A$, if $(A, m) = 1$. (In Table 3.1, $gA = 25 = 3 \cdot 8 + 1$ and $M = 5 \cdot 8 = 40$.) Obtain g and M of the systematic nonseparate codes (if one exists) equivalent to the following AN codes:

   (a) $13N$ code, $0 \leq N < 4$ $(r = 2)$,
   (b) $23N$ code, $0 \leq N < 32$ $(r = 2)$,
   (c) $31N$ code, $0 \leq N < 81$ $(r = 3)$,
   (d) $3N$ code, $0 \leq N < 15$ $(r = 2)$.

2. For a separate adder and checker (Figure 3.2) obtain all permissible values for the check base b which enables the code to be closed under addition and detection of any error $E \in U(m, 2)$ if

(a) $m = 2^6 - 1$,   (b) $m = 2^8 - 1$,   (c) $m = 2^{12}$.

3. Table 3.4 contains formulas for $Q = \Phi_{bi}(N_{1b}, N_{2b})$ for 1's complement case. Obtain a similar table for the 2's complement case. Make sure your formulas are correct by means of examples.

REFERENCES

1. J. L. Diamond, Checking Codes for Digital Computers, *Proc. IRE* **43**, 487–488 (April 1955).
2. D. T. Brown, Error Detecting and Error Correcting Binary Codes for Arithmetic Operations, *IRE Trans. Electron. Comput.* EC-9, 333–337 (September 1960).
3. R. T. Chien, On Linear Residue Codes for Burst Error Correction, *IEEE Trans. Information Theory* IT-10, 127–133 (April 1964).
4. W. W. Peterson and E. J. Weldon Jr., "Error-Correcting Codes," second Ed., The MIT Press, Cambridge, Massachusetts, 1972.
5. H. L. Garner, Error Codes for Arithmetic Operations, *IEEE Trans. Electron. Comput.* EC-15, 763–770 (October 1966).
6. A. Avizienis, Arithmetic Error Codes: Cost and Effectiveness Studies for Application in Digital Systems, *IEEE Trans. Comput.* C-20, 1322–1330 (November 1971).
7. D. S. Henderson, Residue Class Error Checking Codes, *Proc. Nat. Meeting Assoc. Comput. Mach., Los Angeles, California, 1961.* Abstract appears in *Comm. ACM* **4**, 307 (July 1961).
8. J. Rothstein, Residues of Binary Numbers Modulo 3, *IRE Trans. Electron. Comput.* EC-8, 229 (June 1959).
9. W. W. Peterson, On Checking an Adder, *IBM J. Res. Develop.* **2**, 166–168 (April 1958).
10. T. R. N. Rao, Error Checking Logic for Arithmetic Type Operations of a Processor, *IEEE Trans. Electron. Comput.* EC-17, 845–848 (September 1968).
11. S. I. Sitnichenko, Decoding Arithmetic Error Correcting Codes, *Engrg. Cybernetics* 731–736 (1970).

4 SINGLE-ERROR CORRECTION

We discussed single-error detection in Chapter 3. In this chapter, we consider in the first section the theory and implementation of binary single-error-correcting AN codes. In Section 4.2 we consider extension of this theory to nonbinary cases. In Section 4.3 we introduce the cyclic AN codes and derive some useful theory of AN codes, where A equals the product of two primes.

4.1 AN CODES AND PRELIMINARIES

Consider an AN code for $0 \le N < m$ in the radix r representation. There are exactly m distinct codewords (or code numbers) in this code. These are $\{0, A, 2A, 3A, \ldots, (m-1)A\}$. We call m the *range of information*. Let R denote the sum of two codewords AN_1 and AN_2. Then

$$R = AN_1 + AN_2 = A(N_1 + N_2)$$

If N_1 and N_2 are such that $N_1 + N_2 < m$, then the sum is a codeword. But if $N_1 + N_2 \ge m$, then R is not a codeword. In order that the code be closed under addition and also have m distinct codewords, the addition must be defined to obtain

$$R = |AN_1 + AN_2|_M$$

where M equals $A \cdot m$ and is called the *code modulus* or *code range*. Then we also have

$$R = |AN_1 + AN_2|_M = A|N_1 + N_2|_m \qquad (4.1)$$

Thus, the addition of codewords modulo M corresponds to addition of information numbers modulo m. As a sequence of this, the code is now the set of all multiples of A in the ring of integers Z_M. Such a subset in Z_M is called an *ideal* in Z_M. If every element of an ideal is a multiple of one unique element, such as A in this case, then it is called a *principal ideal*, and that unique element is called the *generator* of the ideal. From the algebraic theory of integers, we have that Z_M is a principal ideal ring, since all its ideals are principal ideals and any divisor of M generates an ideal in Z_M.

The complement of AN in Z_M is $M - AN = A(m - N) = (A\overline{N})$, and therefore is a codeword. Also, by definition $W_M(AN) = \min\{W(AN), W(A\overline{N})\}$. The minimum arithmetic weight of the nonzero codewords is also the minimum modular weight. Also for any two codewords AN_1 and AN_2 in Z_M, their difference $|AN_1 - AN_2|_M = A|N_1 - N_2|_m = AN_3$ is also a codeword. The minimum modular distance between two codewords is equal to the modular weight of some other codeword. Therefore the minimum D_M between all pairs of codewords is the minimum W_M of its nonzero codewords.

The minimum modular distance of a code is a very important property of the code and relates very effectively to the error detection and correction properties of the code.

Hereafter we observe the following summary of notation for AN codes in Z_M:

m, the information range, $0 \le N < m$;†
M, the code modulus, $M = A \cdot m$;
$\overline{N}$, the complement of N, $\overline{N} = m - N$;
$\widetilde{AN}$, the complement of AN, $\widetilde{AN} = M - AN$;

---

† In many papers, B instead of m denotes information range. In this book, we preferred to use m to avoid conflict of notation with B of $AN + B$ codes.

$\tilde{E}$, the complement of E, $\tilde{E} = M - E$;

$d_{\min}$, the distance of the code, i.e., the minimum modular distance which also equals the minimum arithmetic distance and the minimum arithmetic weight of the nonzero codewords;

IR, the information rate equal to $\log_r m / \log_r M$.

If the correct result and the actually obtained (possibly erroneous) result of an arithmetic operation on AN codes are denoted by R and R', respectively, then the error E is given by

$$E = |R' - R|_M \tag{4.2}$$

If $W_M(E) = t$, then E is a t-fold error or a pattern of t errors. By the notation introduced in Chapter 2, E is an element of the error sets $U(M, t)$ and $V(M, t)$. E in (4.2) is an element of Z_M and hence is nonnegative, in contrast to the previously defined $E = R' - R$ in the infinite ring of all integers where E can be of either polarity.

Conditions for error detection–correction

For an integer x, let $S(x)$ denote the syndrome of x given by

$$S(x) = |x|_A$$

We assume that the arithmetic operations of interest preserve the code. That is, the result of an arithmetic operation on AN codewords is also a codeword. Therefore the specified result R equals AN for some N. Then the syndrome of the actual result R' is

$$S(R') = |R'|_A = ||R + E|_M|_A = ||AN + E|_M|_A$$

By Lemma 3.2, $||AN + E|_M|_A = |AN + E|_A$, since A divides M, and therefore

$$S(R') = |E|_A = S(E)$$

Thus the syndrome of an arithmetic result is the syndrome of the error E. An error $E \neq 0$ is said to be *detectable under* A iff $S(E) = |E|_A \neq 0$. Therefore we have the following.

LEMMA 4.1

An AN code in Z_M can detect any error in $U(M, d)$ iff for any $E \in U(M, d)$, $S(E) \neq 0$. That is, the error set $U(M, d)$ is detectable under A, the generator of the code, iff the AN code is capable of detecting all errors of modular weight d or less.

Further, if two distinct nonzero errors E_1, E_2 are such that their syndromes $|E_1|_A$ and $|E_2|_A$ are nonzero and unequal, then the errors are said to be *distinguishable under A*. If every distinct pair E_i, $E_j \in U(M, t)$ is distinguishable under A, then the syndromes of all errors in $U(M, t)$ are distinct. Then by association of each of the errors in $U(M, t)$ with a unique syndrome, the exact error can be configured and hence corrected. Therefore we have the following.

LEMMA 4.2

An AN code can correct any error in $U(M, t)$ if every distinct pair in $U(M, t)$ is distinguishable under A.

In other words, there must be a one-to-one map of the error set into (or onto) the set of all syndromes for error correction. The syndrome set, defined as above, is a subset of the integers modulo A.

EXAMPLES

Consider the set of all single errors in the infinite ring of all integers. This set, which consists of all integers of the form $+2^j$ and -2^j, may be denoted by $V(\infty, 1)$. It should be noted that -2^j is the complement of 2^j for this case. This error set is detectable under A, for any odd integer A.

On the other hand, consider the error set $V(M, 1)$ for $M = 2^6 - 1 = 63$. Noting that $V(M, 1) = \{1, 2, 4, 8, 16, 31, 32, 47, 55, 59, 61,$ and $62\}$, this set is detectable under 3, 7, ..., but is not detectable under 5, 11, ...

As an example of error correction, consider $V(65, 1)$ under 13. The one-to-one correspondence between the errors in the set and the syndromes is established by Table 4.1.

Table 4.1

$E \in V(65, 1)$	$S(E) = \lvert E \rvert_{13}$	$\bar{E}$	$S(\bar{E})$
1	1	64	12
2	2	63	11
4	4	61	9
8	8	57	5
16	3	49	10
32	6	33	7

Thus the $13N$ code corrects any error from $V(65, 1)$. Before single-error-correcting codes are derived, we introduce here the relationship between minimum distance (D) of a code and its error detection-correction properties as follows.

THEOREM 4.3　(Massey [1])

An AN code can detect any error pattern E of d or fewer errors [i.e., $W(E) \leq d$] iff the distance (D) between every pair of codewords is at least $d + 1$. Also, an AN code can correct any error pattern E of t or fewer errors [i.e., $W(E) \leq t$] iff the distance (D) between every pair of codewords is at least $2t + 1$.

It should be noted that the weight and distance in Theorem 4.3 are W and D but not W_M and D_M. When W and D are used, the addition structure and the finiteness of the code numbers are implicitly ignored. Use of W_M and D_M, on the other hand, brings in these realistic constraints. Therefore for AN codes in Z_M, an analogous theorem can be

established, provided the code modulus M satisfies the following
inequality for all $N_1, N_2, N_3 \in Z_M$:

$$D_M(N_1, N_2) \leq D_M(N_1, N_3) + D_M(N_3, N_2) \qquad (4.3)\dagger$$

The inequality (4.3) follows readily from the inequality for W_M given
as (2.8). Also by Theorem 2.1, when $M = 2^n$ or $2^n - 1$, the inequality
(2.8) holds and therefore (4.3) holds. This lays the groundwork for the
following.

THEOREM 4.4

Assume that (4.3) holds in Z_M. Then an AN code in Z_M can detect
any error $E \in U(M, d)$ iff the modular distance of the code is at least
$d + 1$. Also, an AN code in Z_M can correct any error $E \in U(M, t)$ iff
the modular distance of the code is at least $2t + 1$.

Proof

We will consider error detection first. Let the modular distance of
the AN code be $d + 1$. Then for any nonzero codeword $AN \in Z_M$,
$W_M(AN) \geq d + 1$. For any error $E \in U(M, d)$, $W_M(E) \leq d$, and if it
is undetectable under A, then $|E|_A = 0$ or $E = Ak$ for some nonzero k,
and therefore E is a codeword and must have $W_M(E) \geq d + 1$. This is a
contradiction, for any $E \in U(M, d)$. Therefore any $E \in U(M, d)$ is detectable.

Conversely assume that any $E \in U(M, d)$ is detectable under A.
Then $|E|_A \neq 0$, which means that E is not a multiple of A and cannot
be a codeword. Since any codeword AN has a syndrome of 0, AN is
not in $U(M, d)$. Since $U(M, d)$ exhausts all integers in Z_M having modular
weight less than or equal to d, except 0, every nonzero codeword
must have a modular weight of at least $d + 1$. This completes the proof
of the first part of the theorem.

---

† When the modular distance satisfies (4.3) in Z_M, then it is a metric in Z_M and we
call Z_M a metric space.

We consider error correction next. Let the modular distance of the code be $2t + 1$. If E_1 and E_2 are two distinct elements of $U(M, t)$, then $W_M(E_i) = W_M(\tilde{E}_i) \leq t$ (for $i = 1, 2$). Assume that E_1 and E_2 are not distinguishable under A. Then $|E_1|_A = |E_2|_A$, and $|E_1 - E_2|_A = 0$. Denoting $E = |E_1 - E_2|_M = |E_1 + \tilde{E}_2|_M$, we have

$$|E|_A = ||E_1 + \tilde{E}_2|_M|_A = |E_1 + \tilde{E}_2|_A = |E_1 - E_2|_A = 0$$

E is therefore a codeword and must have $W_M \geq 2t + 1$. But from the triangular inequality (2.8),

$$W_M(E) = W_M(E_1 + \tilde{E}_2) \leq W_M(E_1) + W_M(\tilde{E}_2) \leq 2t$$

and therefore we have a contradiction. Hence E_1 and E_2 are distinguishable under A and the AN code provides error correction of t or fewer errors.

Conversely, let the set $U(M, t)$ be correctable under A. Also assume that there exists a codeword AX of modular weight less than or equal to $2t$. In its minimal form, AX or its complement $M - AX = \widetilde{AX}$ can be expressed as a polynomial $\sum_{j=0}^{n-1} a_j 2^j$ of $2t$ or fewer nonzero terms ($a_j = 0, 1,$ or -1). By an appropriate separation of the terms of the polynomial into two distinct groups E_1 and E_2, AX or $M - AX$ can be written as the sum $|E_1 + E_2|_M$, such that $W_M(E_i) \leq t$ (for $i = 1, 2$).

Then AX or $\widetilde{AX}$ equals $|E_1 + E_2|_M$. Further,

$$|AX|_A = |\widetilde{AX}|_A = 0 = ||E_1 + E_2|_M|_A = |E_1 + E_2|_A$$

Hence $|E_1 + E_2|_A = 0$ or $|E_1|_A = |\tilde{E}_2|_A$. This means that E_1 and $\tilde{E}_2 \in U(M, t)$ are indistinguishable under A, which contradicts our original assumption. Therefore the code cannot have a nonzero codeword of weight less than or equal to $2t$, as was to be proved   Q.E.D.

More generally, an AN code in Z_M can detect any error in $U(M, d)$ and correct any error in $U(M, t)$ (for $t \leq d$) iff the modular distance of the code is $t + d + 1$. The proof of this is left as an exercise for the reader. (See Problem 4.4.) Table 4.2 lists the choices available for a code with a specified distance.

Table 4.2

Modular distance of the code	Detects $U(M, d)$ and corrects $U(M, t)$		
		d	t
2		1	0
3	(a)	2	0
	(b)	1	1
4	(a)	3	0
	(b)	2	1
5	(a)	4	0
	(b)	3	1
	(c)	2	2
$t + d + 1$		d	t (for all $d \geq t$)

The range of AN codes

Peterson [2] introduced the following notation:

$M_r(A, d)$ is the least positive integer such that the arithmetic weight of $AM_r(A, d)$ in the radix r representation is less than d.

If radix $r = 2$, we denote this integer by $M_2(A, d)$ or simply by $M(A, d)$, without the subscript.†

EXAMPLES

$M(11, 3) = 3$, since $W(11) = W(11 \times 2) = 3$ and $W(11 \times 3) = 2$. Also, $M(3, 2) = \infty$ since $W(3N) \geq 2$ for any arbitrarily large N. $M(13, 3) = 5$ since $W(13N) \geq 3$ for $1 \leq N \leq 4$ and $W(13 \times 5) = 2$.

† The reader should be careful to distinguish between $M(A, d)$ defined here and the modulus M of the AN codes.

Consider an AN code for $0 \le N < m = M_r(A, d)$. Each nonzero codeword has an arithmetic weight of at least d, and the code has a minimum distance of at least d. This minimum distance d cannot be maintained if the range m exceeds $M_r(A, d)$. Thus, determination of $M_r(A, d)$ for given A and d determines the range for the information of an AN code with a minimum distance d. Further, we will show here that $M_r(A, d)$ permits us to determine the range of information for AN codes having a minimum modular distance of d as follows.

Consider an AN code in Z_M, where M as before equals $A \cdot m$ for some $m \le M_r(A, d)$. Then $W(AN) \ge d$ for all $1 \le N < m = M_r(A, d)$ and by definition we have

$$W_M(AN) = \min\{W(AN), W(\widetilde{AN})\}$$
$$= \min\{W(AN), W(A(m - N))\} \ge d$$

Thus the distance of the code is d, and we have

THEOREM 4.5

An AN code in Z_M ($M = A \cdot m$) has a distance d iff $m \le M_r(A, d)$.

For an AN code, the parameter d (the distance) determines the error control properties, whereas M (the code modulus) determines the arithmetic structure of the code. If $M = 2^n - 1$, then the addition can be handled by a conventional binary adder with end-around-carry, and also the complement of AN is obtained by 1's complement of the binary number AN. Further, the code is closed under the operations of addition, complementation, subtraction, and cycle (see Section 4.3).

Given two of the three parameters A, m, and d the coding problem will be to determine the third. As an example, suppose we wish to find a suitable A for a binary code which provides an information range $m = 10$, and a minimum distance of 3. Then we look for an odd integer A, such that $M_2(A, 3) \ge 10$. The smallest such integer is 19 and $M_2(19, 3) = 27$.

Determination of $M(A, 3)$

The problem of finding the range m of information for a given odd integer A to have single-error correction (i.e., a minimum distance of 3) was considered by Brown [3]. Before we state the results of Brown [3] and Peterson [2] on $M(A, 3)$, we introduce the following definitions.

DEFINITION 4.1

The length of r modulo A, for relatively prime r and A, is the least positive integer i such that $r^i \equiv \pm 1 \pmod{A}$. We denote this integer by $L_r(A)$. If the radix $r = 2$, we can drop the subscript for simplicity and write $L(A)$.

DEFINITION 4.2

The order of r modulo A, for relatively prime r and A, is the least positive integer i, such that $r^i \equiv 1 \pmod{A}$. We denote this by $e_r(A)$. If the radix $r = 2$, we can drop the subscript for simplicity and write $e(A)$. The order is sometimes referred to as the *exponent*.

EXAMPLES

$L_2(5) = 2$ since $2^2 \equiv -1 \pmod{5}$; but $e_2(5) = 4$, since the least positive integer i satisfying $2^i \equiv 1 \pmod{5}$ is 4. On the other hand, the length of 2 mod 23, and the order of 2 mod 23 are one and the same. $L(23) = e(23) = 11$. Further, there is no integer i that satisfies $2^i \equiv -1 \pmod{23}$.

In the ring Z_A, the successive powers of r (for relatively prime r and A) generate a multiplicative subgroup in Z_A. As stated earlier in Section 1.2, the set of all integers in Z_A which are relatively prime to A form a multiplicative group, $G(A)$, called the reduced residue system. The multiplicative group generated by r in Z_A is a subgroup of $G(A)$.

If A is a prime, the ring Z_A is a field, and we denote this by $GF(A)$. (Finite fields are called Galois fields, and therefore GF is used.) In that case, $G(A)$ consists of all nonzero elements of $GF(A)$. If r generates the entire multiplicative group $G(A)$, i.e., the successive powers of r exhaust the multiplicative group $G(A)$, then r is said to be a *primitive element* in Z_A. In $GF(A)$, the powers of r exhaust all nonzero elements of $GF(A)$, iff r is a primitive element of $GF(A)$.†

Consider $Z_7 = GF(7)$. The powers of 2 generate $\{2, 4, 8 = 1\}$, and do not exhaust all the nonzero elements; $e_2(7) = 3$. On the other hand, 3 generates $\{3, 3^2 = 2, 3^3 = 6, 3^4 = 4, 3^5 = 5, 3^6 = 1\}$, exhausting the multiplicative group in $GF(7)$, and $e_3(7) = 6$. Hence 3, but not 2, is a primitive element of $GF(7)$. For any primitive element r in $GF(A)$, $e_r(A) = A - 1$.

THEOREM 4.6   (Brown)

Given an odd integer $A \geq 3$, $M_2(A, 3)$ is the integer that satisfies (4.4) for the smallest positive k:

$$M_2(A, 3) = (2^k \pm 1)/A \qquad (4.4)$$

This theorem, in some sense, follows naturally from the definition of $M_r(A, d)$. Since A is odd, AN cannot be equal to 2^k. $M_2(A, 3)$ is the smallest integer N such that the arithmetic weight of AN is 2 or less. Therefore $AM_2(A, 3)$ must have a weight of 2, or in other words, it must be of the form $2^i \pm 2^j$ for $i > j$. By factoring, we obtain $2^i \pm 2^j = (2^{i-j} \pm 1)2^j$. Since 2^j is relatively prime to A, 2^j must divide $M_2(A, 3)$, and for some $N < M_2(A, 3)$, AN is of the form $2^k \pm 1$. This contradicts the definition of $M_2(A, 3)$. Therefore $2^j = 1$ and $AM_2(A, 3)$ must be of the form $2^k \pm 1$.

---

† As a further note on finite fields, there are two types [2, Chapter 6]: (1) $GF(A)$, identically the same as the ring Z_A when A is a prime; (2) $GF(A)$ when A equals p^m for some prime p and $m > 1$. In the latter case $GF(A)$ is not the same as Z_A, but is the set of polynomials with coefficients from $GF(p^m)$ and of degree less than m. In this book we refer only to prime fields, i.e. fields of the first type.

EXAMPLES

Let $A = 19$. Then $L(A) = 9$, since the least positive i satisfying $2^i \equiv \pm 1 \pmod{19}$ is 9. Since $2^9 \equiv -1 \pmod{19}$ we have $M(19, 3) = (2^{L(A)} + 1)/19 = (2^9 + 1)/19 = 27$.

If $A = 23$, then $L(A) = e(A) = 11$ and $M(23, 3) = (2^{11} - 1)/23 = 89$.

If $L(A)$ is known, then $M(A, 3)$ could be obtained from (4.4). The question which then arises is, "How do we find $L(A)$?" We can relate $L(A)$ and $e(A)$ by means of the following.

LEMMA 4.7

Given relatively prime integers r and A, $L_r(A)$ must be either $e_r(A)$ or $e_r(A)/2$.

Proof

From the definitions of $L_r(A)$ and $e_r(A)$, $L_r(A) \le e_r(A)$. Also $r^{L_r(A)} \equiv 1 \pmod{A}$ iff $L_r(A) = e_r(A)$. Assume therefore that $r^{L_r(A)} \equiv -1 \pmod{A}$. Then $e_r(A) \ne L_r(A)$, and

$$(r^{L_r(A)})^2 \equiv (-1)^2 \pmod{A} \quad \text{or} \quad r^{2L_r(A)} \equiv 1 \pmod{A}$$

This means that $e_r(A)$ divides $2L_r(A)$. If $e_r(A) = 2L_r(A)$, then $L_r(A) = e_r(A)/2$, as was to be proved. If $e_r(A)$ is a proper divisor of $2L_r(A)$, then $e_r(A) \le L_r(A)$, which contradicts the two earlier results. This proves the lemma.

By substituting $L(A)$ for k in (4.4), a modified version of Theorem 4.6 is as follows.

THEOREM 4.8

For an odd integer A

$$M(A, 3) = \begin{cases} (2^{L(A)} + 1)/A & \text{iff } 2^{L(A)} \equiv -1 \pmod{A} \\ (2^{L(A)} - 1)/A & \text{iff } 2^{L(A)} = 2^{e(A)} \equiv 1 \pmod{A} \end{cases} \quad (4.5)$$

LEMMA 4.9

r is a primitive element in a prime field $GF(A)$ iff $e_r(A) = A - 1$ and $L_r(A) = (A - 1)/2$.

Proof

Consider $GF(A)$. If r is a primitive element, then $e_r(A) = A - 1$. From Lemma 4.7, $L_r(A) = e_r(A)/2$ or $e_r(A)$. If $L_r(A) = e_r(A)/2$, then $L_r(A) = (A - 1)/2$, as is to be proved. If $L_r(A) = e_r(A)$, then there does not exist a j such that $r^j \equiv -1 \pmod{A}$. Hence r does not generate -1 and therefore this contradicts the assumption that r is a primitive element.† Therefore $L_r(A) = (A - 1)/2$. Conversely, let $e_r(A) = A - 1$. Then the successive powers of r generate $S = \{r, r^2, \ldots, r^{A-1} = 1\}$. The elements of this set S must be distinct since if $r^j = r^i \pmod{A}$ for distinct $j, i \leq A - 1$, then r^{j-i} or $r^{i-j} \equiv 1 \pmod{A}$, which is absurd. Further, r and A are relatively prime and so also are r^j and A for all j. Therefore the set S has $A - 1$ distinct elements, each one relatively prime to A. This can be true only if A is a prime and r is a primitive element of $GF(A)$.

THEOREM 4.10 (Brown and Peterson)

A is an odd prime and 2 is a primitive element of $GF(A)$ iff

$$M_2(A, 3) = (2^{(A-1)/2} + 1)/A \qquad (4.6)$$

Also -2, but not 2, is a primitive element of $GF(A)$ iff

$$M_2(A, 3) = (2^{(A-1)/2} - 1)/A \qquad (4.7)$$

Proof

Let 2 be a primitive element of $GF(A)$ for some prime A. Then from Lemma 4.9, $L(A) = (A - 1)/2$ and $2^{L(A)} \equiv -1 \pmod{A}$. From

---

† For simplicity in $GF(A)$, we use -1 for $A - 1$ and -2 for $A - 2$.

Theorem 4.8, $M_2(A, 3) = (2^{L(A)} + 1)/A$, which establishes (4.6). Conversely, let (4.6) be true. Then $2^{(A-1)/2} \equiv -1 \pmod{A}$ and $L(A) = (A - 1)/2$. From Theorem 4.8 we have $2^{L(A)} \equiv -1 \pmod{A}$. Hence $e(A) = 2L(A) = A - 1$, which, from Lemma 4.9, establishes that 2 is a primitive element of a prime field $GF(A)$. This completes the proof of the first part.

Let -2, but not 2, be a primitive element of $GF(A)$. Then $L_{-2}(A) = (A - 1)/2$ and $e_{-2}(A) = A - 1$. Then $(-2)^{(A-1)/2} \equiv -1 \pmod{A}$. Also, from the definition of $L_r(A)$ it is evident that $L_2(A) = L_{-2}(A)$. Therefore $L_2(A) = (A - 1)/2$ and $2^{(A-1)/2} \equiv \pm 1 \pmod{A}$. Since 2 is not a primitive element, $2^{L(A)} \not\equiv -1 \pmod{A}$. Hence from Theorem 4.8, we have

$$M_2(A, 3) = (2^{L(A)} - 1)/A = (2^{(A-1)/2} - 1)/A$$

as we wished to prove.

Conversely, if $M_2(A, 3) = (2^{(A-1)/2} - 1)/A$, then $L_2(A) = (A - 1)/2 = e(A)$, and 2 is not a primitive element in $G(A)$. Since $L_2(A) = L_{-2}(A)$, $(-2)^{(A-1)/2} \equiv \pm 1 \pmod{A}$. If $(A - 1)/2$ is odd, then $(-2)^{(A-1)/2} = -2^{(A-1)/2} \equiv -1 \pmod{A}$. Therefore -2 is a primitive element in the prime field $GF(A)$, as was to be proved. Consider $|\pm 2^j|_A$ for $j = 0, 1, \ldots, (A - 3)/2$. These residues are distinct and exhaust the nonzero integers modulo A. Hence A is a prime. If $(A - 1)/2$ is even, then $2^{(A-1)/2} - 1 = (2^{(A-1)/4} + 1)(2^{(A-1)/4} - 1) \equiv 0 \pmod{A}$ which means that $L(A) = (A - 1)/4$, which is contradictory. Therefore $(A - 1)/2$ is odd, and the converse holds.   Q.E.D.

The class of codes whose generators A are primes, with 2 or -2 being primitive in $GF(A)$, is called here the Brown–Peterson (BP) codes. The BP codes are single-error correcting ($d_{\min} = 3$); some examples are listed in Table 4.3.

If the modulus of the code $M = A \cdot m = A \cdot M(A, 3)$, then for the codes listed in Table 4.3 we have a one-to-one correspondence between the elements in $V(M, 1)$ and the nonzero elements of A. For instance, consider the $23N$ code $0 \le N < 89$. The number of elements in $V(2^{11} - 1, 1) = 22$, the same as the number of nonzero elements of

Table 4.3 Brown–Peterson codes

A	$L(A)$	$e(A)$	$M(A, 3)$	$A \cdot M(A \cdot 3)$
11	5	10	3	$2^5 + 1$
13	6	12	5	$2^6 + 1$
19	9	18	27	$2^9 + 1$
23	11	11	89	$2^{11} - 1$
29	14	28	565	$2^{14} + 1$
37	18	36	7,085	$2^{18} + 1$
47	23	23	17,841	$2^{23} - 1$
53	26	52	1,266,205	$2^{26} + 1$

$GF(23)$. Since in this code all nonzero elements of $GF(23)$ are utilized for error correction, the code is said to be *perfect* or *close-packed*.

DEFINITION 4.3

An AN code is said to be perfect or close-packed if every element of $G(A)$ is utilized for correction of the errors in the set $U(M, d)$ for some $d \geq 1$.

It can easily be shown that the BP codes are perfect.

Self-complementing $AN + B$ codes†

In the definition of a generalized register which was given earlier, the words "radix" and "base" were used to mean the same thing—the number of distinct symbols in a digit set. Sometimes a distinction between base and radix is made as follows. First consider as an example the representation in binary coded decimal (BCD) number systems. Here a decimal number having a base $b = 10$ has each of its decimal digits represented by a group of binary digits, i.e., of radix 2. Similarly, a ternary-coded, duodecimal system has a radix $r = 3$ and a base $b = 12$, and so on. Table 4.4 contains the BCD systems usually called 8421

† This section is derived largely from the work of Brown [3].

Table 4.4

| N
$0 \le N < b$ | BCD code | | Error-detecting | | Error
correcting |
	8421 code	Excess 3 code $(N + 3)$ code self-complementing	$3N$ code	$3N + 2$ code	$19N + 42$ code
0	0000	0011	00000	00010	00101010
1	0001	0100	00011	00101	00111101
2	0010	0101	00110	01000	01010000
⋮	⋮	⋮	⋮	⋮	⋮
7	0111	1010	10101	10111	10101111
8	1000	1011	11000	11010	11000010
9	1001	1100	11011	11101	11010101

BCD code (or natural BCD code) and excess 3 BCD code as examples. The excess 3 code here is also called the $N + 3$ code, as each N, $0 \le N \le 9$, is represented as $N + 3$ in binary form. The main advantage of the excess 3 code is that the 9's complement of N, which is $9 - N$, is obtained by an exchange of 0's to 1's and 1's to 0's. This is also expressed as $N + 3 + (9 - N + 3) = 2^k - 1$ for some suitable k ($k = 4$ for this case), and $2^k - 1$ has k successive 1's in its binary form. Such codes are called *self-complementing* codes.

Now consider an AN code in a radix r representation, where N is any digit in base b. That is, $0 \le N < b$. If N and its complement (with respect to $b - 1$), $\overline{N} = b - 1 - N$, are to be complement representations in r, then

$$AN + A(b - 1 - N) = r^k - 1 \qquad (4.8)$$

for some suitable k. Often (4.8) cannot be satisfied for given b, r, and A. In such cases a constant B is attached to the AN codeword to make it self-complementing.† Then we have for some suitable k

$$AN + B + (b - 1 - N)A + B = r^k - 1$$
$$(b - 1)A + 2B = r^k - 1 \qquad (4.9)$$

† An $AN + B$ code has the same minimum distance (and therefore the same error control properties) as the AN code since $AN_1 + B - (AN_2 + B) = AN_1 - AN_2$.

As a single-error-detecting BCD code, $0 \leq N < b = 10$, a $3N + B$ code can be investigated. Then from (4.9), $2B = 2^k - 1 - 9 \times 3$, or $B = (2^5 - 1 - 27)/2 = 2$, and therefore the $3N + 2$ code is self-complementing and single-error detecting (Table 4.4). For a single-error-correcting BCD code, we should first look for an A that satisfies $M_2(A, 3) \geq 10$. Since 19 is the smallest prime to satisfy this condition, the $19N + B$ code is a possible candidate. Then $B = (2^k - 1 - 19 \times 9)/2$. This is satisfied for smallest $k = 8$ and $B = (255 - 171)/2 = 42$; therefore we get a $19N + 42$ code. Also $23N + 24$, $25N + 15$, and $27N + 6$ codes are other examples meeting the requirements of self-complementation using no more than eight bits.

The self-complementing codes with $B = 0$ have definite advantages in implementation. For a $7N + B$ code for BCD representation, $B = (2^k - 1 - 9 \times 7)/2$; and if $k = 6$, we have $B = 0$. Thus a $7N$ code, $0 \leq N \leq 9$, is self-complementing (in $r = 2$). Also for a given base b and a radix r, A is chosen to satisfy the error control properties, and B is chosen to make it self-complementing. However, for given b, r, and A, there may not be an integral value for B which satisfies (4.9), and hence an $AN + B$ self-complementing code is not always possible. As an example, a binary-coded quinary code ($b = 5$, $r = 2$) can have $A = 13$ for single-error correction, since $M_2(13, 3) = 5$. Then, from (4.9), $2B = 2^k - 1 - 4 \times 13$ is odd and B has no integral value.

Consider the addition of two codewords $AN_1 + B$ and $AN_2 + B$ for $0 \leq N_i < b$. The actual sum $A(N_1 + N_2) + 2B$ must be corrected (i.e., added to $-B$) to obtain $A(N_1 + N_2) + B$, the code form for the sum $N_1 + N_2 < b$. If $N_1 + N_2 \geq b$, the corrected sum must be $A|N_1 + N_2|_b + B = A(N_1 + N_2 - b) + B$ along with a carry signal advance to the next stage. The correction in this case differs from $-B$. Thus $AN + B$ codes ($B \neq 0$) may have the disadvantage of complicated corrections, which may outweigh the advantages that accrued due to the self-complementing property. In Table 4.4, the $N + 3$, $3N + 2$, and $19N + 42$ codes are self-complementing, as can be readily verified.

Implementing *AN* codes

One possible implementation schematic of an arithmetic processor using *AN*-coded operands is shown in Figure 4.1. The register **A** con-

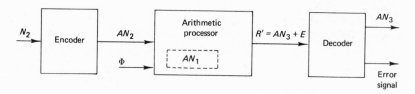

Figure 4.1   A model for an AP using AN-coded operands.

tains the internal operand AN_1 (the codeword for N_1). The other
operand AN_2 and the operation command Φ are applied as inputs to
the processor. The result (possibly erroneous) is $R' = AN_3 + E$, where E
is the error. This result passes through the decoder, which obtains the
syndrome $|R'|_A = |E|_A$, and generates an error signal whenever it is
nonzero. On the other hand, if the syndrome is characteristic of the
error, i.e., if each correctable error has a unique syndrome, then error
correction can be applied. Syndrome generation techniques have been
discussed in Section 3.5. Once the syndrome is generated, the corresponding error E can be obtained by a suitable decoding operation or
by use of a table look-up. This step is called syndrome decoding. The
final step is to correct for the error by subtracting E from R'.

4.2   HIGHER RADIX AN CODES

The theorems on single-error correction by Brown and Peterson are
given for the binary case only. It is commonly felt that these theorems
can be generalized easily to any higher radix. But it has been shown [4]
that such generalizations are not so straightforward. The importance of
higher radix codes stems from the fact that in present-day technology,
groups of binary bits (i.e., bytes) are handled in individual integrated
chips (IC's). This gives a kind of byte organization for processors in
general and arithmetic processors in particular. Consequently, a complex
or multiple error in a byte (or one IC) can affect only one particular
byte. If C is the byte length, any error in one byte is just a single error in

a radix $r = 2^c$ number system. This concept is also appropriate for other radices such as $r = 10$ when binary-coded decimal logic is used. We start here with a generalized version of Brown's theorem (Theorem 4.6) as follows.

THEOREM 4.11   (Rao and Trehan [4])

Let A be greater than $2r$ and $\gcd(A, r) = 1$. Also let N be the smallest integer of the form αr^n for $n > 0$ and $1 \le \alpha < r$, such that

$$|N|_A = |\alpha r^n|_A = \pm \beta \qquad \text{for} \quad 1 \le \beta < r \qquad (4.10)$$

Then

$$M_r(A, 3) = \begin{cases} (\alpha r^n - \beta)/A & \text{if} \quad |\alpha r^n|_A = \beta \\ (\alpha r^n + \beta)/A & \text{if} \quad |\alpha r^n|_A = -\beta \end{cases} \qquad (4.11)$$

Proof

The single-error patterns are of the form $\pm ar^j$ for $j \ge 0$ and $1 \le a < r$. The syndromes of these error patterns can be listed as shown in Table 4.5. For $j = 0$ all the residues are distinct and of the form $|\pm c|_A$ for $1 \le c < r$. These residues are also nonzero, since $\gcd(A, r) = 1$. We would like to show that the entries in the syndrome table (Table 4.5) preceding the entry $|\alpha r^n|_A$ are distinct and nonzero. If this is not so, assume some two entries to be equal, i.e., $|ar^i|_A = |\pm br^k|_A$ for the following two cases:

Case 1: $i, k < n$ and $1 \le a < r$, $1 \le b < r$.
Without loss of generality assume that $i > k$, then

$$ar^i = \pm br^k \pmod{A}$$

or

$$ar^{i-k} = \pm b \pmod{A}$$

This contradicts the hypothesis that N is the smallest integer of the form αr^n such that $\alpha r^n = \pm \beta \pmod{A}$.

Table 4.5 Syndromes of the single-error patterns of the form $|\pm r^j|_A$

j	$|r^j|_A$	$|-r^j|_A$	$|2r^j|_A$	$|-2r^j|_A$	$\cdots$	$|\alpha r^j|_A$	$|-\alpha r^j|_A$	$\cdots$	$|(r-1)r^j|_A$	$|-(r-1)r^j|_A$
0	1	-1	2	-2	$\cdots$	α	$-\alpha$	$\cdots$	$r-1$	$-(r-1)$
1	r	$-r$	$|2r|_A$	$|-2r|_A$	$\cdots$	$|\alpha r|_A$	$|-\alpha r|_A$	$\cdots$	$|(r-1)r|_A$	$|-(r-1)r|_A$
$\cdots$	$\cdots$	$\cdots$	$\cdots$	$\cdots$	$\cdots$	$\cdots$	$\cdots$	$\cdots$	$\cdots$	$\cdots$
$n-1$	$|r^{n-1}|_A$	$|-r^{n-1}|_A$	$|2r^{n-1}|_A$	$|-2r^{n-1}|_A$	$\cdots$	$|\alpha r^{n-1}|_A$	$|-\alpha r^{n-1}|_A$	$\cdots$	$|(r-1)r^{n-1}|_A$	$|-(r-1)r^{n-1}|_A$
n	$|r^n|_A$	$|-r^n|_A$	$|2r^n|_A$	$|-2r^n|_A$	$\cdots$	$|\alpha r^n|_A = \beta$†	$|-\alpha r^n|_A = -\beta$†			

†Indicates repetition.

Case 2: $i = n$, $k \leq n$, and $a < \alpha$.

If
$$ar^n = \pm br^k \pmod{A}$$

we have
$$ar^{n-k} = \pm b \pmod{A}$$

and this again contradicts the hypothesis for the same reason as in Case 1. Hence all the entries in Table 4.5 preceding the entry $|\alpha r^n|_A = \pm \beta$ are distinct and nonzero. Then by Theorem 4.4 we have a minimum weight of 3 for all AN, $0 < N < M_r(A, 3)$, and $M_r(A, 3)$ is given by

$$M_r(A, 3) = \begin{cases} (\alpha r^n - \beta)/A & \text{for} \quad |\alpha r^n|_A = \beta \\ (\alpha r^n + \beta)/A & \text{for} \quad |\alpha r^n|_A = -\beta \end{cases}$$

EXAMPLE

Let $A = 23$, $r = 3$. The syndromes for this case are listed in Table 4.6

Table 4.6

j	$|3^j|_{23}$	$|-3^j|_{23}$	$|+2 \cdot 3^j|_{23}$	$|-2 \cdot 3^j|_{23}$
0	1	$-1 \equiv 22$	2	$-2 \equiv 21$
1	3	$-3 \equiv 20$	6	$-6 \equiv 17$
2	9	$-9 \equiv 14$	18	$-18 \equiv 5$
3	4	$-4 \equiv 19$	8	$-8 \equiv 15$
4	12	$-12 \equiv 11$	1^a	

[a] Indicates repetition.

Hence $\alpha = 2$, $\beta = 1$, giving us

$$M_3(23, 3) = (2 \cdot 3^4 - 1)/23 = 7$$

For single-error correction in an arbitrary radix r, the following theorems of Neumann and Rao [5] are of importance. Let $A = (r - 1)p$ for some prime $p > r$, and let $G_r(p)$ denote the cyclic (multiplicative) subgroup generated by r in $GF(p)$. Also let a^{-1} denote the multiplicative inverse of a in $GF(p)$; i.e., $aa^{-1} \equiv 1 \pmod{p}$.

THEOREM 4.12   (Neumann and Rao [5])

Given that $p - 1$ does not exist in $G_r(p)$ and that the condition

$$(a - r + 1)a^{-1} \notin G_r(p) \qquad \text{for all} \quad a, \qquad 0 < a < r - 1 \qquad (4.12)$$

is satisfied, then

$$M_r(A, 3) = (r^{e_r(p)} - 1)/A \qquad (4.13)$$

Proof

We prove (4.12) using the concept that the AN code is of arithmetic distance (at least) 3 if it is single-error correcting. If it is single-error correcting, then all single errors in radix r have distinct syndromes ar^i (modulo A). We characterize a single error E by the form

$$E = ar^j \qquad \text{for} \quad 0 < |a| < r, \qquad \text{and for all} \quad j = 0, 1, \ldots, e_r(p) - 1$$

Assume that two single errors $E_1 = ar^j$ and $E_2 = br^t$ are such that their syndromes modulo A are equal (and assume that $j \geq t$, without loss of generality). Then

$$ar^j \equiv br^t \pmod{A} \qquad \text{and} \qquad ar^{j-t} \equiv b \pmod{A}$$

From which

$$ar^i \equiv b \pmod{A} \qquad (4.14)$$

for $i = j - t$, some integer among $0, 1, \ldots, e_r(p) - 1$. From (4.14) it follows that

$$ar^i \equiv b \pmod{r - 1} \qquad (4.15)$$

and

$$ar^i \equiv b \pmod{p}, \qquad i = 0, 1, \ldots, e_r(p) - 1 \qquad (4.16)$$

From (4.15) we have that $a \equiv b \pmod{r - 1}$. Thus $a = b$ when a and b are of the same sign. When they are of opposite sign, we can choose a to be positive without loss of generality,

$$a = b + (r - 1) \qquad (4.17)$$

If $a = b$, then (4.16) is contradictory, and therefore the syndromes of all distinct errors must be distinct, as we were to prove. For the case when $a \neq b$, (4.17) holds. Substituting (4.17) into (4.16), we get

$$ar^i \equiv a - (r - 1) \pmod p$$
$$(a - r + 1)a^{-1} \equiv r^i \pmod p, \qquad (a - r + 1)a^{-1} \in G_r(p) \tag{4.18}$$

This is a contradiction of the hypothesis (4.12). Therefore, the theorem is proved. (As an example, the reader might try $r = 8$, $p = 19$, $a = 2$, $i = 2$.)

Next we consider the special case when $-r$ is primitive but r is not primitive in $GF(p)$. We know from number theory that in this case $(-r)^{(p-1)/2} \equiv -1 \pmod p$ and $r^{(p-1)/2} \equiv 1 \pmod p$. Also -1 does not exist in $G_r(p)$, as is required for the hypothesis of Theorem 4.12. This provides us with the following useful result.

THEOREM 4.13

Given that $-r$, but not r, is primitive in $GF(p)$ and that the condition (4.12) is satisfied in $GF(p)$, we have

$$M_r(A, 3) = (r^{(p-1)/2} - 1)/A \tag{4.19}$$

LEMMA 4.14

Given $r = 2^c$ and $p = 2^d - 1$, it follows that condition (4.12) of Theorem 4.12 is satisfied.

Proof

We observe that since p is by definition a prime, d must be a prime. Thus the elements of $G_r(p)$ are precisely the first d consecutive powers of 2. Note that $G_{2^c}(p)$ is identical to $G_2(p)$ whenever $\gcd(c, d) = 1$. Therefore we need only prove that $a \cdot 2^t \equiv a - r + 1 \pmod{2^d - 1}$ cannot occur. Assume that it can. Then

$$a \cdot 2^t \equiv 2^d - 2^c + a \pmod{2^d - 1},$$
$$a \cdot 2^t \equiv 2^c(2^{d-c} - 1) + a \pmod{2^d - 1} \tag{4.20}$$

We note that on the right-hand side of the congruence (4.20) we have an integer less than $2^d - 1$ whose binary representation has two parts, the higher order part of value $2^d - 2^c$ and the lower order part (c digits) of value a. Also we note that the Hamming weight of this integer (the number of 1's in the binary representation) must be at least one greater than the Hamming weight of a. On the other hand, the Hamming weight of $a \cdot 2^t \pmod{2^d - 1}$ is the same as the Hamming weight of a, because the multiplication by a power of 2 modulo $2^d - 1$ is in effect a cyclic shift of a by t places and the Hamming weight is invariant under cyclic shifts. Therefore, the congruence (4.20) cannot hold and the lemma is thus proved. (For a note on Hamming weight, see Section 8.1.)

As a consequence of Lemma 4.14 and Theorem 4.12. we have the following.

THEOREM 4.15

For $r = 2^c$ and $p = 2^d - 1$,

$$M_r(A, 3) = (r^d - 1)/A \tag{4.21}$$

Proof

From Lemma 4.14, condition (4.12) of Theorem 4.12 is satisfied. Further, the elements of $G_r(p)$ are of the form 2^k ($k = 0, 1, \ldots, d - 1$), and -1 clearly is not in $G_r(p)$. Also, since $\gcd(c, d) = 1$, we have $e_r(2^d - 1) = d$. Thus (4.21) follows from Theorem 4.12.   Q.E.D.

Parts of the theorems of Peterson and of Rao and Trehan follow by setting $r = 2$ and $r = 3$, respectively, in Theorem 4.13.

COROLLARY 4.16   (Peterson)†

If -2 but not 2 is primitive in $GF(p)$, then

$$M_2(p, 3) = (2^{(p-1)/2} - 1)/p \tag{4.22}$$

†Corollary 4.16 is only one part of Theorem 4.10.

Proof

In Theorem 4.13, we set $r = 2$, from which $A = p$. We observe that the open interval $(0, r - 1)$ is empty, and therefore (4.12) is trivially satisfied. Then (4.22) follows from (4.19).   Q.E.D.

Note that -2 is a primitive root in $GF(p)$, $p = 8i - 1$ if $4i - 1$ is also a prime [6, Theorem 38].

COROLLARY 4.17   (Rao and Trehan)

Given $A = 2p$, with -3 but not 3 primitive in $GF(p)$, we have

$$M_3(A, 3) = (3^{(p-1)/2} - 1)/A \qquad (4.23)$$

Proof

We set $r = 3$ in Theorem 4.13. There is only one integer in the open interval, $(0, r - 1)$ and that is $a = 1$. For that case we get $(a - r + 1)a^{-1} = 1$, and -1 does not exist in $G_r(p)$ for the same reasons as stated in the proof of Theorem 4.13. Therefore the condition (4.12) is satisfied, and (4.23) follows from (4.19).   Q.E.D.

This sequence of theorems extends to $r = 4$ with a minor modification.

THEOREM 4.18

Given $A = 3p$ with -2 but not 2 primitive in $GF(p)$, we have

$$M_4(A, 3) = (4^{(p-1)/2} - 1)/A \qquad (4.24)$$

Proof

Set $r = 4$ in Theorem 4.13. In the open interval $(0, r - 1)$, there exist only $a = 1$ and $a = 2$, for which $(a - r + 1)a^{-1} = -(3 - a)a^{-1}$ is -2 and $(-1)(-(p - 1)/2) = (p - 1)/2$, respectively. First we note that $-2 \in G_4(p)$ iff $(p - 1)/2 \in G_4(p)$. Assume $-2 \in G_4(p)$. Then $4^a \equiv -2$

(mod p) which means $2^{2a-1} \equiv -1$ (mod p). This would be impossible since -2 is primitive and 2 and is not primitive in $GF(p)$. Therefore condition (4.12) holds, and the theorem is valid.   Q.E.D.

Note that -4 is a primitive root in $GF(p)$, $p = 4i - 1$ if $2i - 1$ is also a prime [6, Theorem 39].

For $r > 4$, the simplicity of the above no longer exists. Condition (4.12) is no longer generally valid, and we must resort to Theorem 4.11.

Theorem 4.13 is thus a generalized form of the Peterson Theorem 4.10 whenever $-r$ but not $+r$ is primitive. Its converse is also true. The original theorems of Peterson and of Rao and Trehan [4] also cover the case of $+r$ primitive for $r = 2$ and 3, respectively, for which cases $p - 1$ is in $G_r(p)$ and

$$M_r(A, 3) = (r^{(p-1)/2} + 1)/A, \qquad r = 2, 3 \tag{4.25}$$

if and only if $+r$ is primitive in $GF(p)$. Unfortunately, (4.25) does not hold for any $r > 3$, since $r - 1$ cannot divide $r^{(p-1)/2} + 1$ for any p. A counterpart of Theorem 4.12 exists in this case, however, as follows.

THEOREM 4.19

Given that $p - 1$ exists in $G_r(p)$ and that condition (4.12) is satisfied in $GF(p)$, then

$$M_r(A, 3) = \begin{cases} (r^{e_r(p)/2} + 1)/p & \text{for even } r \\ (r^{e_r(p)/2} + 1)/2p & \text{for odd } r \end{cases} \tag{4.26}$$

The proof of this has been omitted but is available elsewhere [5].

EXAMPLES   (Neumann–Rao Codes)

$$A = (r - 1)p$$

$r = 2$, Brown–Peterson code, $19N$ code:

$$M_2(19, 3) = (2^9 + 1)/19 = 27$$

$r = 3$, $34N$ code ($p = 17$):

$$M_3(34, 3) = (3^8 + 1)/34 = 193$$

$r = 4$, $69N$ code ($p = 23$) Theorem 4.18:

$$M_4(69, 3) = (4^{11} - 1)/69 = 60787$$

$r = 5$, $44N$ code ($p = 11$) Theorem 4.12:

$$M_5(44, 3) = (5^5 - 1)/44 = 71$$

$r = 8, p = 2^5 - 1, A = 7 \times 31$ (Theorem 4.15):

$$M_8(A, 3) = (8^5 - 1)/A = 151$$

$r = 16, p = 2^7 - 1, A = 15 \times 127$ (Theorem 4.15):

$$M_{16}(A, 3) = (16^7 - 1)/A = 140, 911$$

Theorem 4.13 specifies the existence (or nonexistence) of *near-perfect* single-error-correcting codes. Near-perfect is used here in the sense that essentially all nonzero syndromes except for all multiples of p are used to correct all possible single errors $\pm a$ in each of the n digits of the resulting AN code. Thus the codes covered by (4.22) and by Theorem 4.13 (and its derivatives) are the only near-perfect AN codes for distance 3. (Those for $r = 2$ are of course *perfect*.) Theorem 4.18 indicates that near-perfect codes exist for $r = 4$ whenever -2 but not 2 is primitive in $GF(p)$ (implying that p must necessarily be of the form $8i - 1$). Such codes exist for $p = 7, 23, 47, 71, 79, \ldots$. As further examples, the shortest nontrivial near-perfect codes for $r = 5, 6, 7, 8, 9$, and 10 have $p = 11, 19, 31, 71, 59$, and 31, respectively. The shortest near-perfect code for $r = 16$ has $p = 503$.

4.3   CYCLIC *AN* CODES

We consider here AN codes useful for binary arithmetic operations. If the arithmetic registers (storing AN-coded operands) are of size n bits (or n flip-flops), then each AN codeword is a binary n-tuple, and N will have a range $0 \leq N < m$ where $2^{n-1} \leq A \cdot m = M < 2^n$. *Then the code is said to be of length n.* A cyclic AN code is defined as follows.

DEFINITION 4.4†

An AN code of length n is said to be cyclic iff for any codeword $X = (x_{n-1}, \ldots, x_1, x_0)$, the n – tuple $Y = (x_{n-2}, \ldots, x_0, x_{n-1})$ obtained by shifting cyclically the bits of X to the left once is also a codeword.

THEOREM 4.20

An AN code of length n is cyclic if and only if A generates an ideal in the ring Z_M where $M = 2^n - 1$.

Proof

Let the AN code be cyclic. That is, if $X = (x_{n-1}, x_{n-2}, \ldots, x_0)$ is a codeword, then $X = Ak$ for some k. Also, $Y = (x_{n-2}, x_{n-3}, \ldots, x_0, x_{n-1})$ is a codeword and hence Y is a multiple of A. In other words,

$$X = \sum_{i=0}^{n-1} x_i 2^i = Ak$$

and

$$Y = \sum_{i=0}^{n-2} x_i 2^{i+1} + x_{n-1} = \begin{cases} 2Ak & \text{if } x_{n-1} = 0 \\ 2Ak - (2^n - 1) & \text{if } x_{n-1} = 1 \end{cases}$$

Since Y is a multiple of A in either case, we obtain that A divides $2^n - 1$. It follows from the elementary theory of rings that A generates an ideal in Z_M for $M = 2^n - 1$.

Conversely, if A generates an ideal in Z_{2^n-1}, then A divides $2^n - 1$. Then for all $X = Ak$, a cyclic shift to the right of X gives us a Y such that

$$Y = \begin{cases} 2Ak & \text{for } x_{n-1} = 0 \\ 2Ak - (2^n - 1) & \text{for } x_{n-1} = 1 \end{cases}$$

---

† This definition and the theorem following are very much analogous to the cyclic codes for communication. See, for example, Peterson and Weldon [2, page 207]. The cyclic nature of certain AN codes was first observed by Mandelbaum [7]. Only binary cyclic codes are considered here. These can be readily generalized for any arbitrary r.

In any case A divides Y, and hence Y is a codeword, thus proving that the AN code is cyclic. Q.E.D.

Some examples of cyclic codes of distance 3 are BP codes generated by a prime A such that -2 but not 2, is primitive in $GF(A)$. The $23N$ code of length 11 [for $0 \le N < (2^{11} - 1)/23 = 89$] is such an example. On the other hand, the distance 3 $19N$ code $(0 \le 19N < 2^9 + 1)$ is not cyclic. If one considers the $19N$ code for $0 \le N < (2^{18} - 1)/19$, then that code is cyclic but has a minimum distance of only 2. As cyclic codes of large distance, we discuss in Chapter 6 AN codes for $A = (2^e - 1)/B$, where B is a prime and e is the order of 2 modulo B. These are discussed as *generalized large distance codes* in Chapter 6. If A generates a cyclic code of length n, then A divides $2^n - 1$ and A also divides $2^{jn} - 1$ for any j. That means that there is a cyclic AN code of length jn for all j. However, we observe the convention that unless otherwise stated, the length of an AN cyclic code is the smallest n such that A divides $2^n - 1$. Clearly then n equals $e(A)$.

Consider a cyclic AN code of length n. Then $M = 2^n - 1$. The complement of a codeword AN_1 is denoted $\widetilde{AN_1} = M - AN_1 = A(m - N_1)$ and is a codeword. (Note that $M = A \cdot m$.) The complementation thus consists of simply switching 0's to 1's and vice versa (commonly known as 1's complement operation). Thus, the cyclic AN codes are closed under the operations of addition, complementation, and left (also right) cycle.

4.4 MORE ON $M(A, 3)$

The problem of finding the range of N for a given odd integer A to have single-error correction (i.e., minimum distance 3) was considered in Section 4.1. $M(A, 3)$ for an odd integer A is the least positive integer that equals $(2^k \pm 1)/A$ for some positive k. This leaves the question of the appropriate value of k open, to be solved by investigation. In this respect the results of Henderson [8] and Peterson [2] give an answer when A is a prime and 2 or -2 is a primitive element of $GF(A)$. (See Theorem 4.10.)

We address ourselves here to the question of how to determine $M(A, 3)$ for the case when A is a composite number of the form $A = P_1 P_2$, where P_1 and P_2 are distinct odd primes. Actually we need to determine the value of k in Brown's result, and we will start in this direction as follows.

Let us denote for simplicity the order of $2 \bmod P_i$ by e_i (for $i = 1, 2$). That is, $e_1 = e(P_1)$ and $e_2 = e(P_2)$. Also let (e_1, e_2) denote the greatest common divisor of e_1 and e_2 and $\langle e_1, e_2 \rangle$ denote the least common multiple of e_1 and e_2. We will relate the length of $2 \bmod A$ [i.e., $L(A)$] to e_1 and e_2 as follows.

THEOREM 4.21   (Rao and Garcia [9])

Let $A = P_1 P_2$ for two distinct odd primes P_1 and P_2, and let e_1 and e_2 be orders of 2 modulo P_1 and P_2, respectively. Also let $d = (e_1, e_2)$. Then $L(A) = \langle e_1, e_2 \rangle / 2$ whenever d is even and both e_1/d and e_2/d are odd, or $L(A) = \langle e_1, e_2 \rangle$ otherwise.

Proof

If the order of 2 modulo A is denoted e, then

$$2^e \equiv 1 \pmod{P_1 P_2} \tag{4.27}$$

Further, $2^e \equiv 1 \pmod{P_i}$ for $i = 1$ and 2. Therefore e_i divides e. Since e is the smallest integer satisfying (4.27) we have that

$$e = \langle e_1, e_2 \rangle \tag{4.28}$$

Let us now consider two mutually exclusive possibilities: (1) at least one of e_1 and e_2 is odd, and (2) both e_1 and e_2 are even.

Case 1. Without loss of generality, we assume that e_1 is odd. If $2^{L(A)} \equiv 1 \pmod{A}$, then $L(A) = e(A) = e = \langle e_1, e_2 \rangle$. On the other hand, assume that $2^{L(A)} \equiv -1 \pmod{A}$; then $2^{L(A)} \equiv -1 \pmod{P_1}$. From Lemma 4.7, $L(P_1) = e(P_1) = e_1$ or $e_1/2$. But since $e_1/2$ is not an integer, $L(P_1) = e_1$, and $2^{L(P_1)} \equiv 1 \pmod{P_1}$. Further, in $GF(P_1)$, the multiplicative subgroup generated by 2 cannot have -1, as an element. Therefore $2^{L(A)} \not\equiv -1 \pmod{P_1}$ and $2^{L(A)} \equiv 1 \pmod{A}$, and $L(A) = \langle e_1, e_2 \rangle$.

Case 2. Both e_1 and e_2 are even. From elementary number theory $e = \langle e_1, e_2 \rangle$ can be expressed as

$$e = \begin{cases} (e_1 \cdot e_2)/d = (e_2/d)e_1 = k_1 e_1 & \text{for some } k_1 \quad (4.29) \\ (e_1/d)e_2 \quad = k_2 e_2 & \text{for some } k_2 \quad (4.30) \end{cases}$$

At most one of the k_i is even, and in this case the other must be odd.

First, without loss of generality, consider the case where k_1 is even in (4.29). Assume that $L(A) \neq e(A) = e$. We have from Lemma 4.7, $L(A) = e(A)/2 = e/2$ and

$$2^{e/2} \equiv -1 \pmod{P_1 P_2} \tag{4.31}$$

which implies from (4.29) that

$$2^{e_1 k_1/2} \equiv -1 \pmod{P_1 P_2}$$

and

$$2^{e_1 k_1/2} \equiv -1 \pmod{P_1}$$

This is impossible since k_1 is even and $2^{e_1} \equiv 1 \pmod{P_1}$. Therefore $L(A) = e(A) = \langle e_1, e_2 \rangle$.

Second, if both k_1 and k_2 are odd, then $2^{e_i} \equiv 1 \pmod{P_i}$ and $2^{e_i/2} \equiv -1 \pmod{P_i}$ for $i = 1$ and 2. Then

$$2^{k_i e_i/2} = (2^{e_i/2})^{k_i} \equiv (-1)^{k_i} \pmod{P_i}$$

and therefore $2^{e/2} \equiv -1 \pmod{P_i}$. Using the property of congruence that if $a \equiv b \pmod{P_1}$ and $a \equiv b \pmod{P_2}$ it follows that $a \equiv b \pmod{P_1 P_2}$, we get $2^{e/2} \equiv -1 \pmod{P_1 P_2}$.

This gives us that $L(A) = e/2 = \langle e_1, e_2 \rangle/2$ for the case where e_1/d and e_2/d are both odd, thus completing the proof.

With the help of Theorem 4.21, we can specialize the result of Brown in a way that will be most useful.

COROLLARY 4.22

Let $A = P_1 P_2$ for two odd primes P_1 and P_2; e_1 and e_2 be the orders of 2 modulo P_1 and P_2, respectively; $d = (e_1, e_2)$; and $e = \langle e_1, e_2 \rangle$. Then

$$M(A, 3) = \begin{cases} (2^{L(A)} + 1)/A = (2^{e/2} + 1)/A & \text{if } d \text{ is even and } e_1/d \text{ and} \\ & e_2/d \text{ are both odd} \\ (2^e - 1)/A & \text{otherwise} \end{cases}$$

A generalization of the result of Corollary 4.22 for any arbitrary odd A is due to Kondratyev and Trofimov [10] and is stated here without proof.

THEOREM 4.23   [10]

Let $A = \prod_{i=1} P_i^{\alpha_i}$, $e_i = e(P_i^{\alpha_i})$ for $i = 1,\ 2,\ \ldots,\ s$, and $e = \langle e_1, e_2, \ldots, e_s \rangle$. Then

$$M(A, 3) = \begin{cases} \dfrac{2^{e/2} + 1}{A} & \text{if } e_i \text{ are all even and } \dfrac{e_i}{e_j} = \dfrac{2x + 1}{2y + 1} \\ \dfrac{2^e - 1}{A} & \text{otherwise} \hspace{3.5em} \text{for all } j \neq i \end{cases}$$

This theorem can be proved along the lines of the previous one.

PROBLEMS

1. Establish a syndrome table (such as Table 4.1) for single error correction in the following AN codes

   (a) $29N$ binary code ($M = 2^{14} + 1$)
   (b) $45N$ binary code ($M = 2^{12} - 1$)
   (c) $34N$ ternary code ($M = 3^8 + 1$)
   (d) $75N$ binary code ($0 \leq N \leq 20$)

Also obtain $e_r(A)$, $L_r(A)$, and $M_r(A, 3)$ of the above codes.

2. $G(A)$ denotes the set of all integers in Z_A which are relatively prime to A.

   (a) Show that $G(A)$ forms an abelian group under multiplication.

   (b) Show that the set $\{r, r^2, r^3 \ldots\} = G_r(A)$ in Z_A is also a multiplicative subgroup of $G(A)$, for any r relatively prime to A.

3. (a) If $G_r(A) = G(A)$, then r is said to be a primitive element of $G(A)$. Obtain $G_2(A)$ in $G(41)$ and its cosets in $G(41)$.

   (b) Show that the $41N$ code, $0 \leq N \leq m = 25$ is capable of correcting all burst errors of length 2 or less. (A burst of length 2 or less is of the form E or $3 \cdot E$ where $E \in U(M, 1)$.)

4. Given that (4.3) holds in Z_M, show that an AN code in Z_M is capable of detecting all errors of modular weight d and correcting all errors of modular weight t iff the $d_{\min}$ is $t + d + 1$ $(d \geq t)$.

5. Obtain a shift register schematic with left shift equal to a multiplication by 2 modulo A, where A is

   (a) 19   (b) 23   (c) 47

6. Find for which of the following A Theorem 4.13 applies. Also find $M_r(A, 3)$ for all cases.

   (a) $A = 32$ $(r = 3)$ (b) $A = 22$ $(r = 3)$
   (c) $A = 31 \times 15$ $(r = 16)$ (d) $A = 7 \times 127$ $(r = 8)$

7. By use of Theorem 4.23 obtain $M(A, 3)$ for the following

   (a) $A = 15 \times 7$,   (b) $A = 37 \times 7^2$ (c) $A = 37 \times 19$
   (d) $A = 37 \times 23$

REFERENCES

1. J. L. Massey, Survey of Residue Coding for Arithmetic Errors, *Internat. Comput. Center Bull.* 3 (October 1964).
2. W. W. Peterson and E. J. Weldon, Jr., "Error Correcting Codes," Second Edition. MIT Press, Cambridge, Massachusetts, 1972.

3. D. T. Brown, Error Detecting and Error Correcting Binary Codes for Arithmetic Operations, *IRE Trans. Electron. Comput.* EC-9, 333–337 (September 1960).

4. T. R. N. Rao and A. K. Trehan, Single-Error-Correcting Non-binary Arithmetic Codes, *IEEE Trans. Inform. Theory* IT-16, 604–608 (September 1970).

5. P. G. Neumann and T. R. N. Rao, Error Correction in Byte-organized Arithmetic Processors, *Proc. 1973 Internat. Symp. on Fault-Tolerant Comput.*, San Francisco, California, 1973.

6. D. Shanks, "Solved and Unsolved Problems in Number Theory," Vol. 1. Spartan Books, Washington, D.C., 1962.

7. D. S. Henderson, Residue Class Error Checking Codes, *Proc. Nat. Meeting Assoc.* Computing Machinery, Los Angeles, California, 1961. Abstract appears in *Comm. ACM* 4, 307 (July 1961).

8. D. Mandelbaum, Arithmetic Codes with Large Distance, *JEEE Trans. Information Theory* IT-13 237–242 (April 1967).

9. T. R. N. Rao and O. N. Garcia, Cyclic and Multiresidue Codes for Arithmetic Operations, *IEEE Trans. Inform. Theory* IT-17, 85–91 (January 1971).

10. V. N. Kondratyev and N. N. Trofirnov, Error-correcting Codes with a Peterson Distance not less than Five, *Engrg. Cybernetics* 3, 85–91 (May-June 1969).

5 ERROR CORRECTION USING SEPARATE CODES

In Chapter 4 we considered single-error correction using AN codes. In Chapter 3, we introduced the $[N, |N|_b]$ separate residue codes, and error detection in elementary operations using such codes was discussed. Here we discuss error correction using separate codes.

A generalization of the separate $[N, |N|_b]$ code leads to a multiple-residue (or multiresidue) code as follows. The information number $x \in Z_{m_0}$ is coded in multiresidue form as a $(t + 1)$-tuple X

$$\mathbf{X} = [x, |x|_{m_1}, \ldots, |x|_{m_t}] = [x, x_1, \ldots, x_t]$$

where $x_i = |x|_{m_i}$ is the ith residue check of x modulo m_i and the moduli m_i, for $1 \le i \le t$, are the residue bases. These residue bases are often chosen to be pairwise prime.†

Arithmetic with multiresidue codewords is performed component-wise. Using the symbol $+$ also to denote addition of the codewords, the sum of $\mathbf{X}$ and $\mathbf{Y}$ is

$$\mathbf{X} + \mathbf{Y} = [|x + y|_{m_0}, |x_1 + y_1|_{m_1}, \ldots, |x_t + y_t|_{m_t}]$$

The addition of each of the components is carried out in $(t + 1)$

---

† For improved information rate, the residue bases need to be pairwise prime. But for easy implementation, or for error location, sometimes this constraint is removed.

independent units and the arithmetic is independent in the sense that no carries are transferred from one unit to the next. Therefore, the errors in one unit will not contaminate any of the others. We will call the unit that performs addition modulo m_0 (of the information parts) the *processor* and the one that performs addition modulo m_i the ith *checker* or the *checker* mod m_i.

Separate codes

A possible implementation of AN codes requires: (1) an encoder which forms the codewords AN; (2) an arithmetic unit which performs the operations on the encoded operands; (3) a decoder which, given the result $R' = R + E = \Phi(AN_1, AN_2) + E = AN_3 + E$, finds $|E|_A$ and implements a correction if $|E|_A$ is nonzero and if E is correctable, or perhaps sets up an error control procedure if E is a detectable but non-correctable error. This implementation is shown schematically in Figure 5.1. (For a model of the processor and an explanation of Φ, see Section 2.2.)

The schematic of Figure 5.1 may be capable of controlling errors due to both transient and permanent faults in the arithmetic processor hardware. However, the implementation shown is not only slower in operation when compared with the nonredundant system but may also be very expensive in view of the limited error control provided by the code.

One may alternatively consider an approach, as suggested previously by several researchers, to utilize the arithmetic unit to encode and

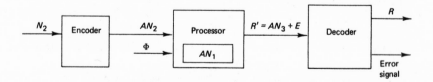

Figure 5.1   Implementation of AN codes.

decode. But this approach suffers from two major drawbacks, namely (1) it can control only transient errors and not permanent errors in the processor, and (2) it slows down system operation considerably.

When compared with the AN codes, the separate codes may have several advantages, as pointed out earlier [1, 2]. These are that (1) the processor and checkers operate independently without propagation of errors from one unit to another; (2) error control of both transient and permanent types is possible; (3) the system operational speed is practically unaffected by error control; (4) the range of the information m can be appreciably increased compared with the AN codes; and (5) the implementation appears relatively easy, particularly if the processor uses 1's complement arithmetic (i.e., $m_0 = 2^n - 1$) and the check bases m_i of the checkers are of the type $2^c - 1$ such that c divides n exactly. A possible organization for error control using a multiresidue code is given later.

5.1 BIRESIDUE CODE

A separate residue code with two checks ($t = 2$) is called a *biresidue code*. For the information $N \in Z_{m_0}$, its codeword is $[N, N_1, N_2]$, where $N_i = |N|_{m_i}$ for $i = 1$ and 2.

DEFINITION 5.1

A triple $[x, y, z]$ is said to be a biresidue codeword with respect to the check bases m_1 and m_2 iff $y = |x|_{m_1}$ and $z = |x|_{m_2}$.

DEFINITION 5.2

A syndrome of a triple $[x, y, z]$ with respect to bases m_1 and m_2, denoted $S[x, y, z]$, is a pair (s_1, s_2), where $s_1 = |x - y|_{m_1}$ and $s_2 = |x - z|_{m_2}$.

From the two definitions just given, we have that a triple $[x, y, z]$ of integers is a biresidue codeword with respect to bases m_1 and m_2 if only its syndrome $S[x, y, z]$ with respect to m_1 and m_2 equals $(0, 0)$.

For the rest of this section, we use the same check bases m_1 and m_2, and codewords and syndromes are assumed to be with respect to these two bases. Further, if two biresidue codewords $[N_1, |N_1|_{m_1}, |N_1|_{m_2}]$, $[N_2, |N_2|_{m_1}, |N_2|_{m_2}]$ are added, we get

$$[|N_1 + N_2|_{m_0}, ||N_1|_{m_1} + |N_2|_{m_1}|_{m_1}, ||N_1|_{m_2} + |N_2|_{m_2}|_{m_2}]$$

Just as for the single residue case $[N, |N|_b]$, the biresidue code will be closed under addition as defined above if and only if the check base m_i divides m_0 for $i = 1$ and 2. Let $[x, y, z]$ be a biresidue codeword. Consider an error E in any one of the components of the codeword as follows.

Case 1   Error E in the processor

The erroneous word in this case is represented by $[x', y, z]$, where $x' = |x + E|_{m_0}$. The syndrome

$$S[x', y, z] = (|x' - y|_{m_1}, |x' - z|_{m_2}) = (s_1, s_2)$$

Further, depending on whether $x + E < m_0$ or $x + E \geq m_0$, we have two cases.

$$|x + E|_{m_0} = \begin{cases} x + E & \text{for} \quad x + E < m_0 \\ x + E - m_0 & \text{for} \quad x + E \geq m_0 \end{cases}$$

Consequently

$$s_1 = ||x + E|_{m_0} - y|_{m_1}$$

$$s_1 = \begin{cases} |E|_{m_1} & \text{if} \quad x + E < m_0 \\ |E - m_0|_{m_1} & \text{if} \quad x + E \geq m_0 \end{cases} \tag{5.1}$$

Similarly,

$$s_2 = \begin{cases} |E|_{m_2} & \text{if} \quad x + E < m_0 \\ |E - m_0|_{m_2} & \text{if} \quad x + E \geq m_0 \end{cases} \tag{5.2}$$

If each of the check bases m_1 and m_2 divides m_0, the range of the information (as required for closure under addition of the codewords), then the equations above reduce to

$$s_i = |E|_{m_i} \quad \text{for} \quad i = 1, 2 \tag{5.3}$$

and the syndrome $S[x', y, z]$ is given by

$$S[x', y, z] = (|E|_{m_1}, |E|_{m_2}) = S(E, 0, 0) \tag{5.4}$$

The error $E \neq 0$ in the processor will not be detected iff

$$|E|_{m_1} = |E|_{m_2} = 0$$

Case 2   Error E in checker 1

The erroneous word here can be denoted by $[x, y', z]$, where $y' = |y + E|_{m_1}$. The syndrome

$$S[x, y', z] = (|x - y'|_{m_1}, |x - z|_{m_2}) = (|-E|_{m_1}, 0)$$

Case 3   Error E in checker 2

The erroneous word here is of the form $[x, y, z']$, where $z' = |z + E|_{m_2}$. The syndrome

$$S[x, y, z'] = (|x - y|_{m_1}, |x - z'|_{m_2}) = (0, |-E|_{m_2})$$

The syndrome of the erroneous word in Cases 2 and 3 has exactly one nonzero component. Further, if the bases m_1 and m_2 are chosen in such a way that for any error in the processor $E \in U(m_0, d)$, $|E|_{m_i} \neq 0$ for $i = 1$ and 2, then the syndrome of the erroneous word has both of its components nonzero. This, in comparison with the syndromes for the Cases 2 and 3, will enable error location as follows. In other words, under the assumption that only one of the three units (the processor, and the checkers 1 and 2) can be erroneous at any given time, the

syndrome of the erroneous word can provide the location of the erroneous unit, based on the following:

  (i) $s_1 \neq 0$, $s_2 \neq 0$: processor error,
  (ii) $s_1 \neq 0$, $s_2 = 0$: checker 1 error,
  (iii) $s_1 = 0$, $s_2 \neq 0$: checker 2 error.

If the error is traced to one of the checkers, then error correction
is available by recomputing the residue from the result of the processor.
On the other hand, if the error is traced to the processor, then the
error value E can be determined from the syndrome (s_1, s_2) if there is
a one-to-one correspondence between the errors of interest and the set
of all possible syndromes with both components nonzero.

Before establishing a correspondence between the error sets in the
processor and the syndromes, the following remarks on the check bases
and information range are in order [3].

(1)   The range of information m_0 equals $2^n - 1$, which also means
that the registers in the processor are of n bits and the logic is 1's
complement arithmetic.

(2)   The check bases m_1 and m_2 are of the form $2^c - 1$ for some
positive integer c. Then the residue generation required for encoding
and decoding will be simple and relatively easy to implement.

(3)   The registers in the checker with base $m_i = 2^c - 1$ are of length
c bits, and further, in order that m_i divide m_0, c must be a divisor
of n. This will provide closure under addition for the code.

The following will provide a basis for the length n of the processor
registers for any two given check bases m_1 and m_2 which are of the
type $2^c - 1$.

LEMMA 5.1

$(a, b) = 1$ iff $(2^a - 1, 2^b - 1) = 1$.

Proof

Let $(2^a - 1, 2^b - 1) = 1$. Assume that $(a, b) = d > 1$. Then from Lemma 3.4, $2^d - 1$ divides $2^a - 1$ and also $2^b - 1$. Since $2^d - 1 > 1$, it follows that $2^d - 1$ divides $(2^a - 1, 2^b - 1)$, which is contradictory. Therefore, our assumption must be wrong and $(a, b) = 1$. Conversely, let $(a, b) = 1$. Assume that $(2^a - 1, 2^b - 1) = d > 1$. Obviously d is odd and d must have a prime $p > 2$ as a factor. Therefore, $2^a - 1 \equiv 2^b - 1 \equiv 0$ (mod p) and $2^a \equiv 2^b \equiv 1$ (mod p). The order of 2 modulo p, denoted $e_2(p)$, divides both a and b. Since $e_2(p)$ is greater than 1 and $e_2(p)$ divides (a, b), we once again arrive at a contradiction. Q.E.D.

If each of the check bases m_1 and m_2 divides the range m_0 exactly, then for any error E in the processor resulting in an erroneous code-word, $[x', y, z]$ satisfies

$$S[x', y, z] = (|E|_{m_1}, |E|_{m_2}) = S[E, 0, 0] \tag{5.5}$$

Further, since any error E in a checker is correctable by the mere location of the erroneous unit, we devote our interest to the more complex task of correcting errors in the processor.

LEMMA 5.2

Let the information range $m_0 = 2^n - 1$, and the check bases $m_1 = 2^a - 1$ and $m_2 = 2^b - 1$. Then the code is closed under addition iff $\langle a, b \rangle$ divides n.

Proof

The code is closed under addition iff each of m_1 and m_2 divides m_0 exactly. From Lemma 3.4, we have that $2^a - 1$ divides $2^n - 1$ iff a divides n. Therefore, a and b are both divisors of n. This means that $\langle a, b \rangle$ is also a divisor of n. Q.E.D.

For closure, n is required to be a multiple of a and b. In addition, for correction of single errors in the processor, n cannot be greater than $\langle a, b \rangle$ due to the following.

THEOREM 5.3

Let $m_0 = 2^n - 1$, $m_1 = 2^a - 1$, and $m_2 = 2^b - 1$. The syndromes corresponding to errors $E \in V(m_0, 1)$ in the processor are distinct if $n = \langle a, b \rangle$, but are not distinct if $n > \langle a, b \rangle$.

Proof

For a codeword $[x, y, z]$, the syndrome $S[x, y, z] = 0$. For an erroneous codeword $[x', y, z]$, where $x' = |x + E|_{m_0}$, the syndrome

$$S[x', y, z] = S[E, 0, 0] = (|E|_{m_1}, |E|_{m_2})$$

Consider the error set $V(m_0, 1)$, which consists of all errors of the type 2^j and $m_0 - 2^j$ for all $j = 0, 1, \ldots, n - 1$. As a first step we will show that $S[2^i, 0, 0] \neq S[m_0 - 2^j, 0, 0]$ for any i, j. As a second step, we will show that for any $i \neq j$ if $S[2^i, 0, 0] = S[2^j, 0, 0]$, then $\langle a, b \rangle$ divides $i - j$. These two steps together establish that the syndromes $S[2^i, 0, 0]$ and $S[m_0 - 2^i, 0, 0]$ for all $i = 0, 1, \ldots, n - 1$ are distinct for $n = \langle a, b \rangle$ but not for $n > \langle a, b \rangle$.

Assume that $S[2^i, 0, 0] = S[m_0 - 2^j, 0, 0]$. Then clearly $|2^i|_{m_1} = |m_0 - 2^j|_{m_1} = |-2^j|_{m_1}$. (Note that m_1 divides m_0.) Therefore, $2^{i-j} \equiv -1 \pmod{m_1}$. The residues modulo $m_1 = 2^a - 1$ of the powers of 2 follow the sequence $1, 2, 4, \ldots, 2^{a-1}, 1, 2, 4, \ldots$, and so on. The only way that one of these residues could be congruent to -1 is that $m_1 = 3$. By the same reasoning, $m_2 = 3$ and $a = b = 2$. $\langle a, b \rangle = 2$ and the two residue checks are the same. Such a code can only correct any error in $V(m_0, 1)$ if $n = m_1$, thus resulting in a trivial code. Hence, for every other case $|2^i|_{m_i} \neq |m_0 - 2^j|_{m_i}$, and therefore

$$S[2^i, 0, 0] \neq S[m_0 - 2^j, 0, 0] \qquad \text{for all} \quad i, j$$

Now let $S[2^i, 0, 0] = S[2^j, 0, 0]$ for $i \neq j$. Then $2^i \equiv 2^j \pmod{m_1}$ or $2^{i-j} \equiv 1 \pmod{m_1}$. Since $2^a \equiv 1 \pmod{m_1}$ for the smallest integer a, we get that a divides $i - j$. Similarly, using base m_2, we get that b divides $i - j$. Therefore, $\langle a, b \rangle$ divides $i - j$. Therefore, $S[2^i, 0, 0] \neq S[2^j, 0, 0]$, provided that $i - j < \langle a, b \rangle$. In other words, the syndromes are distinct for $i = 0, 1, \ldots, n - 1$, provided that $n = \langle a, b \rangle$. By the same argument, if $n > \langle a, b \rangle$, the syndromes cannot be distinct, as was to be proved.

EXAMPLE

Let $m_1 = 2^3 - 1 = 7$, $m_2 = 2^4 - 1 = 15$. From Theorem 5.3 $n = \langle a, b \rangle = \langle 3, 4 \rangle = 12$, and the syndromes corresponding to all single errors are shown in Table 5.1. Note that $|m_0 - 2^i|_{m_i} = |-2^i|_{m_i}$.

Table 5.1

i	Syndrome $S(2^i, 0, 0) = (|2^i|_7, |2^i|_{15})$	Syndrome $S(m_0 - 2^i, 0, 0) = (|-2^i|_7, |-2^i|_{15})$
0	(1, 1)	(6, 14)
1	(2, 2)	(5, 13)
2	(4, 4)	(3, 11)
3	(1, 8)	(6, 7)
4	(2, 1)	(5, 14)
5	(4, 2)	(3, 13)
6	(1, 4)	(6, 11)
7	(2, 8)	(5, 7)
8	(4, 1)	(3, 14)
9	(1, 2)	(6, 13)
10	(2, 4)	(5, 11)
11	(4, 8)	(3, 7)
12	(1, 1)	(6, 14)

It is to be noted that the 24 syndromes listed above the dashed line in Table 5.1 are distinct, and that a repetition occurs for the first time when $i = 12$. The $[N, |N|_7, |N|_{15}]$ code is therefore capable of correcting all single errors in the processor (and, of course, any kind of error in the checkers). If we restrict the moduli m_1 and m_2 to relatively prime integers (of the type $2^c - 1$), then the length n of the processor registers is $\langle a, b \rangle = ab$.

A natural extension of the above leaves us with a multiresidue code of the type $[N, |N|_{m_1}, |N|_{m_2}, \ldots, |N|_{m_t}]$, where $m_i = 2^{a_i} - 1$. For such a code to be single error correcting, the length of the arithmetic register n must be equal to $\langle a_1, a_2, \ldots, a_t \rangle$.

5.2   ERROR CORRECTION USING BIRESIDUE CODES

The method we could employ to achieve error correction is indicated in Figure 5.2. We use two residue checkers as shown. (Checker 1 uses base m_1 and checker 2 uses base m_2.) The arithmetic processor we intend to use as a basis for error control is shown in Figure 5.3. The accumulator register which holds the internal operand N_1 and the augend register

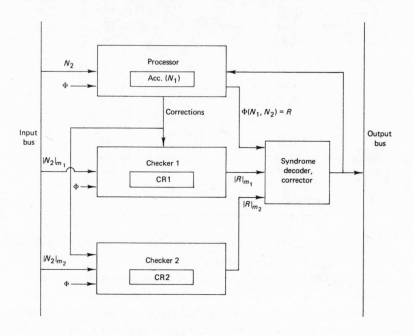

Figure 5.2   Processor and checker organization.

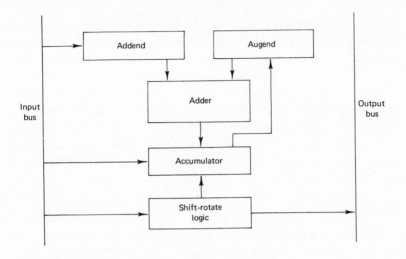

Figure 5.3 Processor organization.

which receives the input operand N_2 are shown. The shift–rotate logic block is indicated in the figure. Checker 1 (also checker 2) in Figures 5.3 and 5.4 contains the check register **CR1** (or **CR2** in the case of checker 2). If R denotes the integer value of the accumulator at the end of any given operation cycle, then **CR1** and **CR2** are to maintain appropriately the corresponding residues $|R|_{m_1}$ and $|R|_{m_2}$, respectively. The three registers, the accumulator, **CR1**, and **CR2** contain a biresidue codeword at the end of each operation cycle.

W consider an operation Φ on the accumulator such as ADD, COM-PLEMENT, SHIFT CYCLE, etc., and indicate the result of such an operation by $R = \Phi(N_1, N_2)$. For each such operation Φ, a pair of corresponding parallel operations Φ_{c1} and Φ_{c2} are to be devised for the registers **CR1** and **CR2**. These parallel operations, which are carried out in the respective residue checkers, will enable maintenance of the correct-residues (and therefore a biresidue codeword) by the end of each operation cycle. However, the errors may be introduced due to logic faults, and these are to be appropriately decoded and corrected. To enable computation in the presence of single errors in the processor (that is, errors of modular weight 1) the syndrome of the erroneous codeword is to be formed and the error value determined. An appropriate correction can be made to the result of the processor. The errors, if traced to a checker, may be left uncorrected, but then the control should prohibit further error correction, although error detection using the healthy checker can be maintained. The procedure to follow when an error is traced to a checker is usually based on the specific application of the computer and the willingness on the part of the designer to expend hardware for correction of errors in the checkers.

The residue checkers

Checkers 1 and 2 are logically identical except that the check bases m_1 and m_2 are different. Each checker consists of the following logic blocks: (1) a residue generator (RG), (2) a residue manipulator (RM), (3) a check register (CR), and (4) a shift–rotate check controller (SRCC). These blocks are shown for the residue checker 1 in Figure 5.4. The residue generator in Figure 5.4 consists of a parallel modulo $2^a - 1$

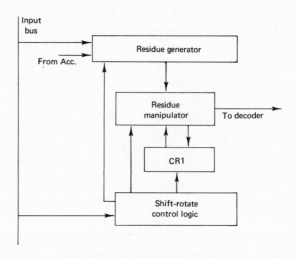

Figure 5.4   Residue checker 1.

adder tree similar to the one discussed in Chapter 3. The residue manipulator is an adder modulo $2^a - 1$, and adds or subtracts (complement and add) the residue obtained from the RG block to the contents of the check register. The logic block SRCC handles ROTATE (i.e., cycle) and CLEAR operations of the CR as required for the various operations, and also generates the control signals to these logic blocks of the residue checker. The residue manipulator generates as output the syndrome s_1 and gates this into the syndrome register 1 (SR1) shown in the syndrome decoder schematic Figure 5.5.

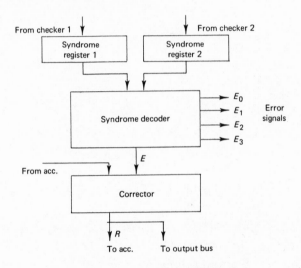

Figure 5.5 Syndrome decoder, corrector.

Syndrome decoder and corrector

This portion of the logic block is shown in Figure 5.5. The syndrome registers SR1 and SR2 receive the generated syndromes s_1 and s_2 from the respective residue checkers. The syndrome decoder determines the erroneous unit, and when the error is traced to the processor, it also determines the error magnitude $|E|$ and the sign of the error. This decoding can be implemented by a memory look-up method or by a hardware logic. A simple hardware syndrome decoder logic useful for biresidue codes using moduli $m_1 = 2^4 - 1$ and $m_2 = 2^5 - 1$ $(m_0 = 2^{20} - 1)$ is illustrated below.

The syndrome decoder (Figure 5.6) generates as outputs the error

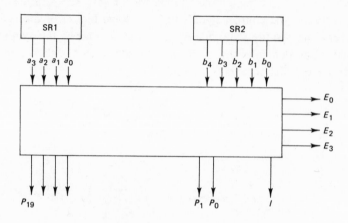

Figure 5.6   Syndrome decoder.

signals E_0, E_1, E_2, E_3, and the error position P_j and polarity indicator I. These outputs represent

$E_0 = 1$:   no error $(s_1 = 0, \quad s_2 = 0)$

$E_1 = 1$:   error in checker 1 $(s_1 \neq 0, \quad s_2 = 0)$

$E_2 = 1$:   error in checker 2 $(s_1 = 0, \quad s_2 \neq 0)$

$E_3 = 1$:   error in the processor $(s_1 \neq 0, \quad s_2 \neq 0)$

The E_i can be readily generated by zero detection signals of SR1 and SR2. Further, when the error is detected in the processor ($E_3 = 1$), the outputs $(P_0, \ldots, P_{n-1})$ and I, the sign indicator represent the error $E = \pm 2^j$ as follows:

$$I = \begin{cases} 0 & \text{if } E \text{ is positive} \\ 1 & \text{if } E \text{ is negative} \end{cases}$$

$$P_j = \begin{cases} 1 & \text{if } |E| = 2^j \\ 0 & \text{otherwise} \end{cases} \quad \text{for } j = 0, 1, 2, \ldots, n-1$$

The sign indicator I is a flip-flop whose value can be generated from the inputs very easily, since the syndrome s_i is of the form 1, 2, 4, 8. When E is positive, only one bit is a 1 in each of the syndrome registers. On the other hand, two or more bits will assume the value 1 in each SR when E is negative, and therefore can be detected easily. When E is negative, I is set to 1 and the syndrome registers are complemented. Otherwise, I is reset to 0 and the syndrome registers are left unchanged. Then the generation of P_j is given by the functions

$$P_0 = a_0 b_0$$
$$P_1 = a_1 b_1$$
$$P_2 = a_2 b_2$$
$$P_3 = a_3 b_3$$
$$P_4 = a_0 b_4$$
$$P_5 = a_1 b_0$$
$$\vdots$$
$$P_{19} = a_3 b_4$$
$$P_j = a_l b_k \qquad \text{where} \quad l = |j|_3, \qquad k = |j|_4$$

The error E in the processor is characterized by the outputs P_{19}, $\dots$, P_0 and I of the syndrome decoder as follows:

$$E = (-1)^I \sum_{i=0}^{19} P_i 2^i \tag{5.6}$$

The error correction once again may be implemented in a number of ways as discussed by Garcia [4]. Here we give a correction schematic using an iterative circuit approach. This correction circuit is a ripple-carry (or borrow) adder–subtractor and uses 20 identical cells (or modules). A typical cell j is shown in Figure 5.7. Let $R' = (r'_{19}, r'_{18}, \dots, r'_1, r'_0)$ denote the actual result (to be corrected), $R = (r_{19}, \dots, r_0)$ the corrected result, and let E be defined by the P_i's and I as in (5.6). The inputs to the jth cell are r'_j, P_j, I, and C_j, and the outputs r_j and C_{j+1} are given by

$$r_j = r'_j \oplus (P_j + C_j), \qquad j = 0, 1, \dots, 19$$
$$C_{j+1} = P_j \bar{r}'_j I + P_j r'_j \bar{I}, \qquad C_{20} = C_0$$

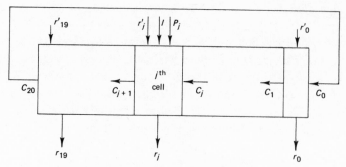

Figure 5.7 Error corrector schematic.

This cell is somewhat more complex than a half-adder due to the add–subtract feature, but is less complex than a full adder.

5.3 CONSTRUCTION OF SEPARATE CODES FROM NONSEPARATE CODES†

Here we derive results which allow the construction of multiresidue codes corresponding to a given AN code in such a way that under some natural constraints the error-correcting properties are alike. The computer organizations necessary for separate and nonseparate codes are quite different from each other. It is therefore to be expected that the error control properties of corresponding codes have to be considered in the light of the particular organization used. This is done by means of appropriate hypotheses (or assumptions) to be established at first with regards to the nature of errors and where they occur. In the organization for separate codes two possible categories of errors are analyzed, and later these two categories are combined into a working hypothesis. Finally examples are discussed to illustrate this theory and the advantages of the separate code organization.

The implementation of AN codes is shown in Figure 5.1, and has been discussed in earlier chapters. Here we consider a schematic for implementing multiresidue codes. The theory and application of the

† This section draws heavily from the work of Garcia [4].

Implementation of separate codes

systematic codes of the nonseparate category, namely $|gAN|_M$ codes, are considered in Chapter 7.

Figure 5.8 describes a schematic for a separate code with t residue checkers, i.e., for a $[N, |N|_{m_1}, |N|_{m_2}, \ldots, |N|_{m_t}]$ code. The information

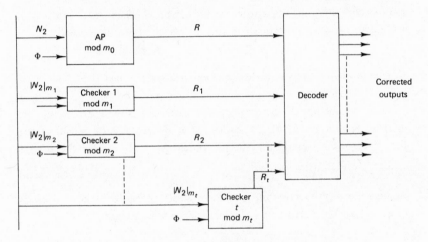

Figure 5.8 Implementation of a multiresidue code.

part N of the codeword is processed in the arithmetic unit, whereas the ith residue $|N|_{m_i}$ ($i = 1, 2, \ldots, t$) is processed in parallel in the ith checker. The arithmetic performed is modulo m_0 for the processor and modulo m_i for the ith checker ($i = 1, 2, \ldots, t$). The approach here is a straightforward generalization of the biresidue code implementation discussed in the last section (Figure 5.2).

Errors in separate codes and categories of errors

In AN codes it is impossible to distinguish information bits from the check bits, and therefore it is meaningless to consider whether the errors occur in the information or in the checks. This is not so in the systematic codes. Particularly in separate codes, it is of fundamental importance as to whether the error occurs in the processor or in the checkers.

We distinguish two categories of errors: (1) errors in the processor, denoted by the set $U(m_0, d)$ for some d, and (2) errors in one or more checkers.

The following hypothesis is valid for the rest of this section: Errors may occur either in the processor or in the checkers, but errors will not occur in both the processor and the checkers simultaneously.

Correspondence between AN codes and separate codes

We can establish an important relation between errors in the arithmetic modulo M of AN codes ($M = A \cdot m$) and the errors of the first category in separate codes in which $m_0 = A \cdot m$ and the m_i are pairwise relatively prime factors of $A = \prod_{i=1}^{t} m_i$. We follow this result by considering the possibility of distinguishing between the two kinds of errors and of correcting errors of the second category.

1. Errors in the processor

There is a natural correspondence between errors in a nonseparate AN code and errors only in the processor of a multiresidue code as specified below. First we define the syndrome of a separately coded word $\mathbf{X} = (x, x_1, x_2, \ldots, x_t)$ by the t-tuple

$$S[\mathbf{X}] = (s_1, s_2, \ldots, s_t)$$

where

$$s_i = |x - x_i|_{m_i} \quad \text{for} \quad i = 1, 2, \ldots, t$$

Let $\mathbf{Z} = (z, z_1, z_2, \ldots, z_t)$ be the correct result of the addition of two multiresidue codewords. If an error E has occurred in the processor, then clearly

$$S[\mathbf{Z} + E] = S[|z + E|_{m_0}, z_1, \ldots, z_t]$$
$$= (|E|_{m_1}, |E|_{m_2}, \ldots, |E|_{m_t})$$

LEMMA 5.4

Let $A = \prod_{i=1}^{t} m_i$, where the m_i are pairwise relatively prime. Given an AN code capable of correcting all $E = U(M, d)$ for $M = A \cdot m$, there exists a separate (t-residue) code with t checkers having bases $m_1, m_2, \ldots, m_t$, and the range of the processor $m_0 = M$ such that there exists a distinct syndrome for each $E \in U(m_0, d)$ in the processor. Conversely, if we are given a multiple-residue code with distinct syndromes corresponding to each $E \in U(m_0, d)$ in the processor, then there exists a corresponding AN code that has the same error-correcting properties.

Proof

The correspondence between AN codes and the multiresidue codes is as follows. Any error E in AN codes has a syndrome $|E|_A$. Any error E in the processor part of the multiresidue code has a syndrome

$$S[E] = (|E|_{m_1}, \ldots, |E|_{m_t})$$

since each m_i divides A and therefore divides m_0. If the syndromes of all $E \in U(M, d)$ are distinct, then they are distinct elements of the ring of integers modulo A. From Theorem 1.18, therefore, the syndromes of all errors $E \in U(m_0, d)$ in the processor are also distinct. The converse follows similarly.

2. Errors in the checkers

Let $\mathbf{X} = (x, x_1, \ldots, x_t)$ be a multiresidue codeword, the syndrome $S[\mathbf{X}] = (0, \ldots, 0)$. An error in the checker i results in an erroneous codeword $\mathbf{X}' = (x, x_1, \ldots, |x_i + E|_{m_i}, \ldots, x_t)$. Obviously the syndrome of $\mathbf{X}'$ is all 0's, except in the ith component of the syndrome, which is $|-E|_{m_i}$. Similarly, if some s checkers are in error and the processor is perfect, then the syndrome of the result of an arithmetic operation will have s nonzero components.

3. Errors in the processor or the checkers

We consider errors of either category, but not both. That is, if errors occur in the adder, then they do not occur in the checkers, and vice versa. Consider a multiresidue code $(N, |N|_{m_1}, \ldots, |N|_{m_t})$, where $0 \leq N < m_0$, and m_i divides m_0 for all $i = 1, 2, \ldots, t$. Assume that at least s out of the t checkers can independently detect all errors $E \in U(m_0, d)$ in the adder. Then any error $E \in U(m_0, d)$ in the adder results in a syndrome that has at least s nonzero components. Since errors belong to the adder or the checkers but not both, erroneous unit location is as follows.

If syndrome has l nonzero components, then we have

$$
\begin{aligned}
&l = 0: && \text{no errors at all} \\
&0 < l < s: && l \text{ checkers in error, the erroneous} \\
& && \text{checkers correspond to the} \hspace{2em} (5.7) \\
& && \text{nonzero components of the} \\
& && \text{syndrome} \\
&l \geq s: && \text{error } E \in U(m_0, d) \text{ in the adder}
\end{aligned}
$$

If errors are traced to the checkers, then they may be corrected by a generation of the correct residues from the information (i.e., processor component). Next, let $m_0 = M$ for some M, and let there be an AN code capable of correcting all errors $E \in U(M, d)$. If $A = \langle m_1, m_2, \ldots, m_t \rangle$ such that for any $E \in U(M, d)$, $|E|_{m_i} \neq 0$ for at least some s check bases, then we can construct a multiresidue code $(N, |N|_{m_1}, \ldots, |N|_{m_t})$ for $0 \leq N < m_0 = M$ with the following properties. Any errors $E \in U(m_0, d)$ in the processor will result in a unique syndrome that also has at least s nonzero components. Hence error unit location based on (5.7) is useful. By virtue of Lemma 5.4, any error in the adder $E \in U(m_0, d)$ is correctable. Also errors in $(s - 1)$ or fewer checkers are correctable. Hence we have the following.

THEOREM 5.5

Consider an AN code with $d_{\min} = d + c + 1$, $d \geq c$. Also let $A = \langle m_1, m_2, \ldots, m_t \rangle$ such that for any $E \in U(M, d)$, $|E|_{m_i} \neq 0$ for at least some s check bases. Then the multiresidue code with t check bases

$m_1, m_2, \ldots, m_t$ and $m_0 = AM(A, d + c + 1)$ is capable of correcting all $E \in U(m_0, c)$ and detecting all $E \in U(m_0, d)$ in the processor or any errors in $(s - 1)$ or fewer checkers.

Applications and examples

From our previous theoretical discussion for the AN codes where $A = p_1 p_2$ (Section 4.4), we can find $M(A, 3)$ from e_1 and e_2. This presents an immediate way of constructing single-error-correcting biresidue codes. Given that $p_1 N$ and $p_2 N$ codes are both single-error detecting, we can actually correct any single error $E \in U(m_0, 1)$ in the adder, where $m_0 = AM(A, 3)$, or any error in one of the checkers. Further, when we consider the practical implementation of these codes it is advantageous to have moduli of the form $2^c - 1$, and also $m_0 = 2^n - 1$.

As an example let

$$p_1 = 15, \qquad p_2 = 127$$
$$e_1 = 4, \qquad e_2 = 7$$
$$d = \gcd(4, 7) = 1$$

From Theorem 4.23, $L(A) = \langle e_1, e_2 \rangle = 28$, and $M(A, 3) = (2^{28} - 1)/A$ $= (2^{28} - 1)/(15 \times 127)$. Therefore $m_0 = 2^{28} - 1$. The biresidue code $(N, |N|_{15}, |N|_{127})$ will correct any single error $E \in U(2^{28} - 1, 1)$ in the adder or any error in either checker. Note here that any single error in the processor is detected by both checkers. (This follows from Theorem 5.3 as well.)

When p_1 and p_2 are not of the form $2^c - 1$, the matter of facility of implementation is open to question. Using the theory, however, we can determine their error-correcting properties. For example, the $(N, |N|_3, |N|_{29})$ code is capable of single-error correction in a 28-bit adder or any error in one of the two checkers. Notice that $e(3) = 2$, $e(29) = 28$, and $L(A) = e(A) = \langle 28, 2 \rangle = 28$. In a biresidue code the range for information is $m_0 = 2^{28} - 1$. One may compare the information rates for the two biresidue code examples as against the possible implementation techniques.

For codes of large distance we can use the results obtained on AN codes by Mandelbaum [5] and Barrows [6] for values of A of the type

$$A = (2^{B-1} - 1)/B$$

where B is a prime, with 2 as a primitive element of $GF(B)$. For example, a distance 6 code is given by $A = (2^{18} - 1)/19 = (3^3)(7)(73)$, but as is characteristic of these large-distance AN codes, the range for N is very limited, $0 \le N < 19$. When this code is implemented as a multiresidue code in an appropriate way we can correct errors $E \in U(m_0, 2)$ in the processor under the assumption that no errors occur in the checkers, but now the range for information is $0 \le N < 2^{18} - 1$. If the errors can occur in the checkers also, then the equivalent multiresidue code $(N, |N|_{27}, |N|_{511})$ can correct only $E = V(m_0, 1)$ in the adder, or any error in one checker. A double-error-correcting biresidue code with m_1 and m_2 not pairwise relatively prime is possible as follows.

EXAMPLES

Let $m_1 = 27 \times 7$, $m_2 = 73 \times 3$, $m_0 = 2^{18} - 1$, and $A = \text{lcm}$ $\langle m_1, m_2 \rangle$. Each residue $|E|_A$ can be represented by a unique pair $(|E|_{m_1}, |E|_{m_2})$ since $A = \text{lcm}\langle m_1, m_2 \rangle$. For any double error E in the processor, i.e., $E \in U(2^{18} - 1, 2)$, $|E|_{m_i} \ne 0$ for $i = 1, 2$, and therefore each checker independently detects E. Also since $|E|_A$ is distinct for each $E \in U(2^{18} - 1, 2)$ the syndromes $(|E|_{m_1}, |E|_{m_2})$ for the set of errors $U(2^{18} - 1, 2)$ are distinct. Hence the biresidue code is capable of correcting all double errors in $U(m_0, 2)$ in the processor or any error in one checker.

Consider an extension to the above-mentioned code as follows. Let $m_0 = 2^{18} - 1$, $m_1 = 27 \times 7$, $m_2 = 73 \times 3$, and $m_3 = 73 \times 3$ for a triresidue code $(N, |N|_{m_1}, |N|_{m_2}, |N|_{m_3})$. The checkers 2 and 3 are identical and are therefore duplicates. $\langle m_1, m_2, m_3 \rangle$ still equals $A = (2^{18} - 1)/19$, and for each $E \in U(m_0, 2)$ in the adder, its syndrome will have three nonzero components. This code, therefore, can correct any error $E \in U(m_0, 2)$ in the adder or any errors in two or fewer checkers.

Other known results on multiresidue codes are reported by Garcia [4], Dadayev [7, 8], and Monteiro and Rao [9].

Garcia approached the problem of correcting errors on the basis of the Hamming weight (the number of nonzero components) of the syndrome, and his results are extensions of those given by Theorem 5.5. Dadayev [7] considered correction of multiple errors (say, t errors) randomly distributed over the entire codeword, which includes the information and $2t$ residue checks. That is, there was no distinction made between errors in the processor and those in the checkers. The conditions he derived for the check bases and for t error correction are too numerous and are very difficult to satisfy.

Monteiro and Rao [9] have derived a class of multiresidue codes with three check bases of the form $2^c - 1$. These codes can correct any two errors and detect any three errors in the processor or any error in one checker. These have been derived from the class of an AN code where $A = \prod_{i=1}^{r} (2^{a_i} - 1)$ and are discussed as "codes with distance not less than five" by Kondratyev and Trofimov [10]. For complete details the reader is advised to refer to these papers.

PROBLEMS

1. Consider the cyclic AN code where

$$A = (2^{B-1} - 1)/B$$

for any prime B, with 2 or -2 primitive in $GF(B)$. This code is known to be of length $B - 1$ and has

$$d_{\min} = \left[\frac{B + 1}{3} \right]$$

(This result is proved in Chapter 6.) Use the above to obtain two biresidue codes $(N, |N|_{m_1}, |N|_{m_2})$ with information range $m_0 = 2^{B-1} - 1$ and $A = m_1 \cdot m_2$ for relatively prime check bases for each of the following cases. Also obtain their error control properties.

(a) $B = 13$.   (b) $B = 23$.   (c) $B = 19$.

2. For the problem above, delete the constraint $(m_1, m_2) = 1$. Then $A = \langle m_1, m_2 \rangle$. Obtain all possible biresidue codes which could correct all errors in $U(m_0, 2)$ or any error in one checker.

3. Consider a multiresidue code with t checkers as follows: The information range $m_0 = 2^{18} - 1$ and the code can correct any error $E \in U(m_0, 2)$ in the processor or any error in at most two checkers.

   (a) What is the minimum number of checkers required? What conditions must be fulfilled by the check bases?

   (b) Obtain two different sets of check bases which satisfy the error control properties.

REFERENCES

1. T. R. N. Rao, Error Checking Logic for Arithmetic Type Operations of a Processor, *IEEE Trans. Comput.* EC-17, 845–848 (September 1968).

2. T. R. N. Rao, Biresidue Error Correcting Codes for Computer Arithmetic. *IEEE Trans. Comput.* C-19, No. 5, (May 1970).

3. T. R. N. Rao and O. N. Garcia, Cyclic and Multi-Residue Codes for Arithmetic Operations, *IEEE Trans. Information Theory* IT-17, 85–91 (January 1971).

4. O. N. Garcia, Error Codes for Arithmetic and Logical Operations, Ph.D. Thesis, Dept. of Elec. Engrg., Univ. of Maryland, College Park, June 1969.

5. D. Mandelbaum, Arithmetic Codes with Large Distance, *IEEE Trans. Information Theory* IT-13, 237–242 (April 1967).

6. J. T. Barrows, Jr., "A New Method for Constructing Multiple Error Correcting Linear Residue Codes," Rep. R-277. Coordinated Sci. Lab., Univ. of Illinois, Urbana, January 1966.

7. Yu. G. Dadayev, Arithmetic Divisible Codes with Correction for Independent Errors, *Engrg. Cybernetics* 79–88 (November-December 1965).

8. Yu. G. Dadayev, Arifmeticheskiye Kody Ispra-Vlyayushehiye Oshibki (Arithmetic Codes that Correct Errors), *Soviet Radio* (1966).

9. P. M. Monteiro and T. R. N. Rao, Multi-Residue Codes for Double Error Correction, *Proc. IEEE-TCCA Symp. Comput. Arithmetic, Dept. of Elec. Engrg. Univ. of Maryland, College Park, May 1972.*

10. V. N. Kondratyev and N. N. Trofimov, Error-Correcting Codes with Peterson Distance Not Less Than Five, *Engrg. Cybernetics* 85–91 (May-June 1969).

6 LARGE-DISTANCE CODES

The *AN* codes with distance greater than 3 are sometimes called *large-distance codes*. Before these codes are introduced, we study algorithms for conversion of binary numbers to other forms such as *modified binary* and *nonadjacent* forms. These algorithms are useful for determining the distance of the class of codes whose generator A is of the form $(2^{e(B)} - 1)/B$ for some odd integer B.

6.1 ALGORITHMS

We define modified binary form (MBF) and nonadjacent form (NAF) for N as follows.

DEFINITION 6.1

If an m-tuple $(a_{m-1}, \ldots, a_0)$ for N satisfies

$$N = \sum_{i=0}^{m=1} a_i 2^i, \qquad a_i = 0, 1, \text{ or } -1 \qquad (6.1)$$

then it is called a modified binary form (MBF) of N.

It is very simple to observe that for a given $N \neq 0$, there are multiple MBF forms. However, $N = 0$ has a unique MBF that is obviously given by all $a_i = 0$.

DEFINITION 6.2

If $(a_{m-1}, \ldots, a_0)$, an MBF of N, satisfies

$$a_i a_{i+1} = 0 \qquad \text{for} \quad i = 0, 1, \ldots, m - 2 \qquad (6.2)$$

then it is said to be a nonadjacent form (NAF) of N.

It is important to note that for a given N, its NAF is unique. Further, the NAF for N has the least number of nonzero coefficients (a_i's) among all MBF's of N [1]. Therefore the NAF is also a minimal form of N. As stated in Chapter 2, the number of nonzero coefficients in the NAF of N represents the arithmetic weight $W(N)$. Therefore, given $(a_{m-1}, \ldots, a_0)$, the NAF of N, we have

$$W(N) = \sum_{i=0}^{m-1} |a_i| \qquad (6.3)$$

The uniqueness of NAF can be proved as follows. Assume that there are two distinct NAF's for N, namely, $(a_{m-1}, \ldots, a_0)$ and $(b_{n-1}, \ldots, b_0)$. The difference between these two forms gives us $N - N = 0 = \sum (a_i - b_i)2^i$. With possible carry or borrow propagations, this will result in an MBF with some nonzero coefficients for 0, which is absurd. Therefore $a_i = b_i$ for all i and the NAF form is unique.

Starting with an MBF for N, a simple algorithm can be given to convert it to an NAF. This algorithm, given next, transforms every pair of nonzero coefficients to a new form with nonadjacency. This transformation would often decrease the total number of nonzero coefficients, but in no case would it increase them. This combined with the fact that the NAF is unique proves that it is a minimal MBF form. The reader should note, however, that there may be other MBF's for N which are also minimal.

ALGORITHM 6.1

Given $(b_{m-1}, \ldots, b_0)$, an MBF of N, the following rules can be applied to convert it to an NAF.

Rule 1

Starting from b_0, proceed to higher order coefficients until a coefficient b_k is reached such that $b_k \cdot b_{k+1} \neq 0$. That is, the coefficients are in NAF up to the coefficient b_{k-1}. The coefficients b_k, b_{k+1} satisfy one of the following four cases. An appropriate transformation is made as follows.

Rule 2

(a) $b_{k+1} = 1, b_k = 1$: set $b_{k+2} = b_{k+2} + 1, b_{k+1} = 0, b_k = -1$.
(b) $b_{k+1} = 1, b_k = -1$: set $b_{k+1} = 0, b_k = 1$.
(c) $b_{k+1} = -1, b_k = 1$: set $b_{k+1} = 0, b_k = -1$.
(d) $b_{k+1} = -1, b_k = -1$: set $b_{k+2} = b_{k+2} - 1, b_{k+1} = 0, b_k = 1$.

The above sets $b_{k+1} = 0$ in all cases, and therefore the NAF is extended up to b_{k+1}. However, b_{k+2} can now be 2, -2, 1, -1, or 0. If b_{k+2} is 1, -1, or 0, the algorithm can now proceed from b_{k+1} using Rule 2. If $b_{k+1} = 2$ or -2, Rule 3 can be applied as follows.

Rule 3

If $b_k = 2$, set $b_{k+1} = b_{k+1} + 1$, $b_k = 0$, or if $b_k = -2$, set $b_{k+1} = b_{k+1} - 1, b_k = 0$.

Continuing these rules for higher order coefficients, an NAF can be obtained.

EXAMPLE

$N = 1 \ 11 \ \bar{1}1 \ 01 \ \bar{1}0 \ 10$ ($\bar{1}$ represents -1).

By Rule 2b for coefficients b_3 and b_4, we get

$$N = 1 \ 11 \ \bar{1}1 \ 00 \ 10 \ 10$$

By Rule 2c for coefficients b_6 and b_7, we get

$$N = 1 \ 11 \ 0\bar{1} \ 00 \ 10 \ 10$$

By Rule 2a for coefficients b_8 and b_9, we get

$$N = 2 \ 0\bar{1} \ 0\bar{1} \ 00 \ 10 \ 10$$

By Rule 3a for coefficients b_{10}, we finally obtain NAF

$$N = 10 \ 0\bar{1} \ 0\bar{1} \ 00 \ 10 \ 10$$

The NAF conversion thus may increase the number of coefficients by at most one [2].

ALGORITHM 6.2

For integers N and B ($N < B$), the following can be used to convert the fraction N/B into a binary sequence:

$$N/B = 0 \cdot q_1 q_2 \cdots q_m \cdots$$

Let $r_0 = N$. We obtain q_i's by using iteratively

$$2r_i = q_{i+1}B + r_{i+1}, \qquad i = 0, 1, 2, \ldots$$

where

$$q_{i+1} = \begin{cases} 1 & \text{if} \quad 2r_i \geq B \\ 0 & \text{if} \quad 2r_i < B \end{cases} \qquad (6.4)$$

The algorithm is terminated when $r_i = 0$.

It should be noted that Algorithm 6.2 may not necessarily terminate, and in fact, may yield a sequence such as $q_1 q_2 \cdots q_j, \ q_{j+1}q_{j+2} \cdots q_{j+p}q_{j+1}q_{j+2} \cdots$, which is recurring with some period p, unless $N/B = N'/2^n$. Also for any given integer N, by choosing $B = 2^n > N$, the algorithm just given can be used to convert N to its binary form.

ALGORITHM 6.3

The following algorithm due to Chang and Tsao-Wu [3] is given here without proof. Starting from the binary forms N and $3N$, we obtain the NAF of N as follows.

$$3N = \sum_{i=0}^{m+1} a_i 2^i, \qquad a_i = 0 \quad \text{or} \quad 1$$

and

$$N = \sum_{i=0}^{m-1} b_i 2^i, \qquad b_i = 0 \quad \text{or} \quad 1$$

Then the NAF of N is given by $(C_{m+1}, C_m, \ldots, C_0)$ where

$$C_i = a_{i+1} - b_{i+1}, \qquad i = 0, \ldots, m+1 \tag{6.5}$$

EXAMPLE

$$
\begin{aligned}
3N = 81 &= 1\ 01\ 00\ 01 \\
N = 27 &= \underline{\quad 1 \quad 10\ 11} \\
\text{NAF of } 2N &= \overline{\quad 1\ 00\ \bar{1}0\ \bar{1}0} \\
\text{NAF of } N &= \quad 10\ 0\bar{1}\ 0\bar{1}
\end{aligned}
$$

It is interesting to note that for an arbitrary radix $r > 2$, NAF forms do not exist for all integers. However Clark and Liang [4] show that there exists for each N a unique modified radix-r form which is also a minimal weight form. They also show that this minimal form is obtained by the digitwise subtraction of the radix-r form of N from the radix-r form of $(r + 1)N$. This method is clearly a generalization of the Algorithm 6.3.

EXAMPLE

Minimal form of $N = 199$ (radix $r = 10$)

$$
\begin{aligned}
11N &= 2189 \\
N &= \ 199 \\
10N &= 20\bar{1}0 \\
\text{minimal form of } N &= \ 20\bar{1}
\end{aligned}
$$

ALGORITHM 6.4

An algorithm, appearing in a paper by Chien *et al.* [5] and Goto and Fukumura [6] provides a direct conversion of a fraction N/B into an NAF as follows. Given integers N and B, B odd and $N < B$, set $r_{-1} = N/2$ and compute the sequence $q_0, q_1, \ldots$ iteratively, using

$$2r_i = q_{i+1}B + r_{i+1} \quad \text{for} \quad i = -1, 0, 1, 2, \ldots \quad (6.6)$$

where

$$q_{i+1} = \begin{cases} 1 & \text{if} \quad 2r_i > \tfrac{2}{3}B \\ 0 & \text{if} \quad \tfrac{2}{3}B \geq 2r_i > -\tfrac{2}{3}B \\ -1 & \text{if} \quad 2r_i \leq -\tfrac{2}{3}B. \end{cases} \quad (6.7)$$

This algorithm yields

$$N/B = q_0 \cdot q_1 q_2 \cdots$$

The following theorem due to Chien *et al.* [5] establishes the validity of the conversion algorithm.

THEOREM 6.1

By Algorithm 6.4 the sequences $q_1 q_2 \cdots$ and $r_1 r_2 \cdots$ satisfy the following:

$$2B/3 \geq r_i > -2B/3 \qquad \text{for all} \quad i \qquad (6.8)$$

$$r_i \equiv N \cdot 2^i \pmod{B} \qquad \text{for all} \quad i \qquad (6.9)$$

$$q_i q_{i+1} = 0 \qquad \text{for all} \quad i \qquad (6.10)$$

Proof

By our original assumption of $N < B$, $r_{-1} = N/2$, and r_{-1} satisfies (6.8). From (6.6), $r_0 = N - q_0 \cdot B$. If $q_0 = 0$, then from (6.7), r_0 must satisfy (6.8). If $q_0 = 1$, once again r_0 satisfies (6.8). Assume now that r_i satisfies (6.8). From Equations (6.6) and (6.7), when $q_{i+1} = 0$, we have $2r_i = r_{i+1}$ and r_{i+1} satisfies (6.8). When $q_{i+1} \neq 0$, $r_{i+1} = 2r_i - q_{i+1} \cdot B$, and once again r_{i+1} must satisfy (6.8). Thus, (6.8) is proved by induction.

We note that $r_0 \equiv 2r_{-1} = N$ (mod B) and from (6.6), $r_1 \equiv 2r_0 \equiv 2N$ (mod B) and $r_2 \equiv 2r_1 \equiv 2^2 N$ (mod B). By repeated use of (6.6) we arrive at

$$r_i \equiv 2r_{i-1} \equiv 2^2 r_{i-2} \cdots \equiv 2^i r_0 \equiv 2^i \cdot N \pmod{B}$$

This proves (6.9).

Finally, to prove (6.10), we start with case $i = 0$. If $q_0 = 0$, (6.10) is clearly satisfied. If $q_0 = 1$, then from (6.6), $r_1 = 2r_0 - B = N - B$. From (6.7), $2r_0 = N < 2B/3$, and therefore $0 < r_1 < 2B/3 - B = -B/3$. Further, $0 > 2r_i > -2B/3$, and from (6.7) we have $q_1 = 0$, as required. If $q_i = 0$, (6.10) is satisfied. If $q_i \neq 0$, assume that $q_i = 1$. (The case $q_i = -1$ can be considered in a similar manner.) We have from (6.8) that $2B/3 > r_{i-1} \geq B/3$ and $4B/3 > 2r_{i-1} \geq 2B/3$. Since $q_i = 1$, $r_i = 2r_{i-1} - B$ and we get $B/3 > r_i \geq -B/3$. This means, from (6.7), that $q_{i+1} = 0$ and therefore satisfies (6.10).

LEMMA 6.2

If $0 < N < B/3$ in Algorithm 6.4, then $q_0 = q_1 = 0$ and

$$q_{i+e(B)} = q_i \qquad \text{for all} \quad i \tag{6.11}$$

Proof

From (6.6) and (6.7) it follows easily that $q_0 = 0$ and $0 < r_0 = N < B/3$. This implies that $0 \leq 2r_0 = r_1 < 2B/3$ and $q_1 = 0$, as we wished to prove. To prove (6.11), we first note that $r_i \equiv 2^i \cdot N$ (mod B) and $2B/3 \geq r_i > -2B/3$, from Theorem 6.1. Therefore, $r_{e(B)} \equiv 2^{e(B)} \cdot N$ (mod B). Since $2^{e(B)} \equiv 1$ (mod B), we get $r_{e(B)} \equiv N$ (mod B). Since $0 < N < B/3$ and $2B/3 \geq r_{e(B)} > -2B/3$, we get $r_{e(B)} = N = r_0$, and as a consequence $r_{e(B)+1} = r_1$ and $r_{e(B)+i} = r_i$. Therefore we get $q_{e(B)+i} = q_i$ for all i, as was to be proved. Q.E.D.

It can be easily shown that the sequence $q_0 \cdot q_1 q_2 \cdots$ obtained for N/B by Algorithm 6.4 is periodic with a period $e(B)$ after the occurrence of the first nonzero q_j. That is, $q_i = q_{i+e(B)}$ for all $i > j$ when q_j is the first nonzero term of the sequence. However, when $N < B/3$, Lemma 6.2 provides that the sequence is periodic with period $e(B)$ right from the start. We make the following definition for the recurring subsequence as follows.

DEFINITION 6.3

For $N/B = q_0 \cdot q_1 q_2 \cdots$ obtained by Algorithm 6.4, let $q_{i+e(B)} = q_i$ for all $i > j$ for the smallest j. Then the subsequence $q_{j+1} q_{j+2} \cdots q_{j+e(B)}$ is said to be the period of N/B.†

From Lemma 6.2, we have that if $N < B/3$, the period of N/B is $q_1 \cdots q_{e(B)}$.

EXAMPLES

Let $N = 8$ and $B = 11$. From Algorithm 6.2, we get the following sequence:

$$
\begin{aligned}
2N = 2r_0 &= 16 = 1 \times 11 + 5, & q_1 &= 1 \\
2r_1 &= 10 = 0 \times 11 + 10, & q_2 &= 0 \\
2r_2 &= 20 = 1 \times 11 + 9, & q_3 &= 1 \\
2r_3 &= 18 = 1 \times 11 + 7, & q_4 &= 1 \\
2r_4 &= 14 = 1 \times 11 + 3, & q_5 &= 1 \\
2r_5 &= 6 = 0 \times 11 + 6, & q_6 &= 0 \\
2r_6 &= 12 = 1 \times 11 + 1, & q_7 &= 1 \\
2r_7 &= 2 = 0 \times 11 + 2, & q_8 &= 0 \\
2r_8 &= 4 = 0 \times 11 + 4, & q_9 &= 0 \\
2r_9 &= 8 = 0 \times 11 + 8, & q_{10} &= 0 \\
\hline
2r_{10} &= 16 = 1 \times 11 + 5, & q_{11} &= 1
\end{aligned}
$$

$$N/B = 0.1011101000 | 1011101000 | \cdots$$

which clearly is a recurring sequence with period $e(11) = 10$.

We observe that $q_{i+10} = q_i$ and $q_{i+5} = 1 - q_i$ for all $i > 0$ and that the period of $N/B = 1011101000$.

---

† Chien et al. [5] call this the B-period of N/B.

From the direct NAF conversion algorithm 6.4 we get

$$N = 2r_{-1} = \quad 8 = \quad 1 \times 11 - 3, \qquad q_0 = \quad 1$$

$$
\begin{aligned}
2r_0 &= -6 = & 0 \times 11 - 6, & \quad q_1 &= \ 0 \\
2r_1 &= -12 = & -1 \times 11 - 1, & \quad q_2 &= -1 \\
2r_3 &= -4 = & 0 \times 11 - 4, & \quad q_4 &= \ 0 \\
2r_4 &= -8 = & -1 \times 11 + 3, & \quad q_5 &= -1 \\
2r_5 &= \ 6 = & 0 \times 11 + 6, & \quad q_6 &= \ 0 \\
2r_6 &= 12 = & 1 \times 11 + 1, & \quad q_7 &= \ 1 \\
2r_7 &= \ 2 = & 0 \times 11 + 2, & \quad q_8 &= \ 0 \\
2r_8 &= \ 4 = & 0 \times 11 + 4, & \quad q_9 &= \ 0 \\
2r_9 &= \ 8 = & 1 \times 11 - 3, & \quad q_{10} &= \ 1
\end{aligned}
$$

(6.12)

$$2r_{10} = -6 = \quad 0 \times 11 - 6, \qquad q_{11} = 0$$
$$\vdots$$

$$N/B = 8/11 = 1 \cdot |0\bar{1}00\bar{1}01001|0\bar{1}00\bar{1}01001| \cdots$$

This sequence also has the period $e(11) = 10$. It can be observed that $q_{i+10} = q_i$ and $q_{i+5} = -q_i$.

6.2 BARROWS–MANDELBAUM (BM) CODES

In Section 4.3, we defined cyclic codes as ideals in the ring Z_{2^n-1}. We observed then that when A is a prime with 2 or -2 as a primitive element in $GF(A)$, A generates a perfect single-error-correcting code with code modulus $M = 2^{(A-1)/2} \pm 1$. Such a code, called the Brown–Peterson (BP) code, is analogous to the Hamming cycle code, which is an ideal in the algebra of polynomials modulo $X^n - 1$, and which is generated by a primitive polynomial. This is a remarkable analogy between the arithmetic codes and communication codes as first pointed out by Massey [2]. There are other analogies and parallels one could draw between these two classes of codes. The Barrows–Mandelbaum codes we study in this chapter are the dual codes of the Brown–Peterson codes in the sense that the generator A of one becomes the information range m of the other (its dual). Therefore, the Barrows–Mandelbaum code is analogous to the maximal length sequence code, whose generator $g(x) = (x^n - 1)/p(x)$, where $p(x)$ is a primitive polynomial having n as the order of its roots.

The cyclic AN code generated by

$$A = (2^{B-1} - 1)/B \tag{6.13}$$

where B is a prime with 2 or -2 primitive in $GF(B)$, has been studied independently by Barrows [7] and Mandelbaum [8]. The range for N is $0 \le N < m = B$ and the length of the code is $B - 1$.† The BM codes obviously have a small information range and therefore the rate of information,

$$IR = \frac{\log_2 B}{B - 1} \tag{6.14}$$

is rather poor. But the minimum distance of the BM codes is very good, as is shown below.

First consider an odd prime B, with 2 as primitive in $GF(B)$. Then from Algorithm 6.4 and Lemma 6.2, we obtain

$$1/B = 0 \cdot q_1 q_2 \cdots q_{e(B)} | q_1 q_2 \cdots q_{e(B)} | \cdots \tag{6.15}$$

Multiplying (6.15) by $2^{e(B)}$ and subtracting (6.15) from it, we get

$$A = (2^{e(B)} - 1)/B = q_1 q_2 \cdots q_{e(B)} = \text{the period of } 1/B \tag{6.16}$$

Clearly the period of $1/B$ is the NAF representation for A. The arithmetic weight of A is given by the number of nonzero terms of (6.16). From (6.9), we have

$$r_i \equiv 2^i \pmod{B}$$

and the r_i's exhaust the complete system of residues modulo B, due to

† The length of a cyclic code is often defined as the smallest n for which A divides $2^n - 1$. When 2 (-2 but not 2) is primitive in $GF(B)$, the length of the BM code is $B - 1$ (($B - 1)/2$). In either case we consider here BM codes of length $B - 1$.

the fact that 2 is primitive in $GF(B)$. We can account for the nonzero q_i's in (6.16) from (6.7), which can be more appropriately written as

$$
\begin{aligned}
q_{i+1} &= 0 && \text{if} && |r_i|_B \leq B/3, && |r_i|_B > 2B/3 \\
q_{i+1} &\neq 0 && \text{if} && 2B/3 \geq |r_i|_B > B/3
\end{aligned}
\qquad (6.17)\dagger
$$

The numbers of nonzero q_i's are exactly as many integers of Z_B which fall in the semiclosed interval $(B/3, 2B/3]$. For convenience we call this interval the middle-third of Z_B. Noting that the middle-third of Z_B contains $[(B + 1)/3]$ integers, we get $W(A) = [(B + 1)/3]$.

Next we note that for any nonzero $N \in Z_B$, $N = |2^i|_B$ for some i. Denoting $2^{e(B)} - 1 = M$, we obtain

$$
AN = \frac{2^{e(B)} - 1}{B} \cdot N = \frac{M}{B} \cdot N = \frac{M}{B} |2^i|_B = \left| \frac{M}{B} \cdot 2^i \right|_M \qquad (6.18)
$$

$$
= |A \cdot 2^i|_M
$$

Therefore AN can be obtained by cyclic shifting of A by i places to the left. (Note that one cyclic shift to the left amounts to a multiplication by 2 modulo M.) By the cyclic shifting, the leading nonzero coefficient could be -1, in which case AN is represented by the negative quantity $AN - M$. However, the reader should note that $AN - M = -(M - AN) = -\widetilde{AN}$ and the complement of AN has "the" minimal form (NAF). (See the next example.) The modular weight of AN therefore is the same as the modular weight of A. This completes the proof of the following.

THEOREM 6.3

The distance (and modular distance) of the cyclic code generated by

$$
A = (2^{e(B)} - 1)/B
$$

for a prime B having 2 as primitive in $GF(B)$ is exactly $[(B + 1)/3]$.

† Since B is a prime, $B/3$ and $2B/3$ are never exact integers except in the trivial case when $B = 3$.

EXAMPLE

$B = 11$, 2 is primitive in $GF(11)$, and $1/11 = 0.0010\bar{1}00\bar{1}010010\bar{1}00\bar{1}01$
$\ldots$. $A = (2^{10} - 1)/11 = 0010\bar{1}00\bar{1}01$, and $M = 2^{10} - 1$. The NAF forms
of AN or $AN - M = -\widetilde{AN}$ for all $N \in (0, B)$ are as follows:

$$
\begin{aligned}
A &= 0010\bar{1}00\bar{1}01 &&= A \\
2A &= 010\bar{1}00\bar{1}010 &&= 2A \\
3A &= |2^8 A|_M = 010010\bar{1}00\bar{1} &&= 3A \\
4A &= |2^2 A|_M = 10\bar{1}00\bar{1}0100 &&= 4A \\
5A &= |2^4 A|_M, \quad \bar{1}00\bar{1}010010 &&= 5A - M \\
6A &= |2^9 A|_M = 10010\bar{1}00\bar{1}0 &&= 6A \\
7A &= |2^7 A|_M, \quad \bar{1}010010\bar{1}00 &&= 7A - M \\
8A &= |2^3 A|_M, \quad 0\bar{1}00\bar{1}01001 &&= 8A - M \\
9A &= |2^6 A|_M, \quad 0\bar{1}01001010 &&= 9A - M \\
10A &= |2^5 A|_M, \quad 00\bar{1}010010\bar{1} &&= 10A - M
\end{aligned}
$$

THEOREM 6.4

Consider a prime B having -2, but not 2, as primitive in $GF(B)$.
Then $e(B) = (B - 1)/2$ and the AN code generated by

$$A = (2^{e(B)} - 1)/B = (2^{(B-1)/2} - 1)/B \qquad (6.19)$$

is cyclic with code length $e(B) = (B - 1)/2$.

For this code $A = (2^{e(B)} - 1)/B = q_1 q_2 \cdots q_{e(B)}$ as before, and the
weight of A is given by the number of integers of the form $|2^i|_B$ falling
in the middle-third of Z_B given by (6.17). We observe that the multiplicative subgroup generated by 2, namely $G_2(B)$ in $GF(B)$, has only
$(B - 1)/2$ integers and of these, those falling in the interval $(B/3, 2B/3]$
contribute to the weight of A. This number can be determined as follows.
For each $N = |2^i|_B$ in the interval $(B/3, 2B/3]$, $B - N = |-2^i|_B$ is also
in the same interval. It can be seen from the results of Chapter 4, that
$-1 \notin G_2(B)$ and that the coset of $G_2(B)$ containing -1 has integers of

the form $|-2^i|_B$. Since the integers of the form $|2^i|_B$ and $|-2^i|_B$ exhaust the nonzero integers of $GF(B)$, we have that the integers of $G_2(B)$ falling in the interval $(B/3, 2B/3]$ are exactly half as many as in the previous case. Therefore the weight of A equals $[(B + 1)/6]$.

Next to obtain the distance of the code, we observe that for $N \in Z_B$, $N \neq 0$, $N = |2^i|_B$ or $|-2^i|_B$, for some i. If $N = |2^i|_B$, AN is obtained by a cyclic shift of A to the left by i places. As before, we note that the weight does not alter by cyclic shifts. Also $-A \equiv (B - 1)A \pmod{M}$ has the same weight as A, and therefore when $M = |-2^i|_B$, AN is obtained by cyclic shifts of the NAF of $-A$. This concludes the proof of the following.

THEOREM 6.5

If B is a prime having -2, but not 2, as primitive in $GF(B)$, then the cyclic AN code generated by

$$A = (2^{e(B)} - 1)/B \tag{6.20}$$

is of length $e(B) = (B - 1)/2$, and has distance (and modular distance) given by

$$d_{\min} = [(B + 1)/6] \tag{6.21}$$

It will be of some interest to note that for B as in Theorem 6.5, a cyclic code of length $B - 1$ can be obtained by using the generator A'

$$A' = (2^{B-1} - 1)/B = ((2^{e(B)} - 1)/B) \cdot (2^{e(B)} + 1) = A(2^{e(B)} + 1)$$

This code is twice as long, with the second half of the digits a mere repetition of the first half. The $d_{\min}$ is also doubled by this duplication. In Table 6.1, we list some distance 2 (all of them cyclic), distance 3 (some cyclic), and the corresponding BM codes of length $e(B)$. The BM codes and the BP codes (called the dual codes) are such that the generator of one code equals the information range of the other (its dual). These two classes occupy the two extremes, one having low $d_{\min}$ and large range m and the other having large $d_{\min}$ but very small m. The

Table 6.1 BP and BM codes.

Distance 2 cyclic codes of smallest length				Distance 3 BP codes				BM cyclic codes, length $n = B - 1$ or $(B - 1)/2$			
A	m	n	$IR = k/n$	A	m	n	k/n	A	$m = B$	k/n	$d_{\min}$
11	$\dfrac{2^{10} - 1}{11}$	10	·653	11^a	$\dfrac{2^5 + 1}{11} = 3$	5	·316	$\dfrac{2^{10} - 1}{11}$	11	·347	4
13	$\dfrac{2^{12} - 1}{13}$	12	·685	13^a	$\dfrac{2^6 + 1}{13} = 5$	6	·385	$\dfrac{2^{12} - 1}{13}$	13	·315	4
19	$\dfrac{2^{18} - 1}{19}$	18	·764	19^a	$\dfrac{2^9 + 1}{19} = 27$	9	·528	$\dfrac{2^{18} - 1}{19}$	19	·236	6
23	$\dfrac{2^{22} - 1}{23}$	22	·795	23	$\dfrac{2^{11} - 1}{23}$	11	·590	$\dfrac{2^{11} - 1}{23}$	23	·410	4
29	$\dfrac{2^{28} - 1}{29}$	28	·824	29^a	$\dfrac{2^{14} + 1}{29}$	14	·648	$\dfrac{2^{28} - 1}{29}$	29	·176	10
37	$\dfrac{2^{36} - 1}{37}$	36	·855	36^a	$\dfrac{2^{18} + 1}{37}$	18	·710	$\dfrac{2^{36} - 1}{37}$	37	·145	12
47	$\dfrac{2^{46} - 1}{47}$	46	·880	47	$\dfrac{2^{23} - 1}{47}$	23	·759	$\dfrac{2^{23} - 1}{47}$	47	·241	8
53	$\dfrac{2^{52} - 1}{53}$	52	·890	53^a	$\dfrac{2^{26} + 1}{53}$	26	·780	$\dfrac{2^{52} - 1}{53}$	53	·110	17
61	$\dfrac{2^{60} - 1}{61}$	60	·901	61^a	$\dfrac{2^{30} + 1}{61}$	30	·802	$\dfrac{2^{60} - 1}{61}$	61	·099	20

^{a} Noncyclic codes.

codes that lie between these two extremes having moderate d_{min} and good information rate are of considerable interest. Chien *et al.* [5] have discussed such codes, and their results are summarized in the following section.

6.3 CHIEN–HONG–PREPARATA (CHP) CODES

From here on B represents an odd integer and $A = (2^{e(B)} - 1)/B$ represents the generator of a cyclic AN code. This code has information range $m = B$ and code modulus $M = A \cdot B = 2^{e(B)} - 1$. First consider $N < B/3$. From Algorithm 6.4, we obtain the sequence for $N/B = 0 \cdot q_1 \cdots q_{e(B)}q_1 \cdots q_{e(B)} \cdots$. Since $N < B/3$ we have from Lemma 6.2 that $q_0 = q_1 = 0$. As before, we obtain $AN = (2^{e(B)} - 1)N/B = q_1 \cdots q_{e(B)}$ as its NAF. The weight of AN is the weight of the period of N/B. Consider N even in $(0, B/3)$. $N' = N/2$ is also in $(0, B/3)$. The weight of the B period of N/B is simply a cyclic shift of the period of N'/B, and therefore AN and AN' have the same weights.

The complement of AN is $\widetilde{AN} = M - AN$, and its weight can be shown to be no less than the weight of AN. Therefore the modular weight of AN equals the modular weight of $A\bar{N}$, which equals the weight of the period N/B. Note that $\bar{N}$ is in $(2B/3, B)$ for any $N \in (0, B/3)$.

For N in $[B/3, 2B/3]$, we consider two separate cases: (1) N is even, and (2) N is odd. When N is even, $N = 2N'$ and N' then is in the interval $(0, B/3)$. Note that in the period of N'/B, $q_1 = 0$, $AN = 2AN'$ and AN is a cyclic shift of AN', and therefore they have the same modular weight. Finally for odd N in $[B/3, 2B/3]$, its complement $B - N$ is even and is also in the same interval. Therefore their modular weights are the same. This completes the proof of the following.

THEOREM 6.6

The cyclic AN code where $A = (2^{e(B)} - 1)/B$ for odd B has modular distance equal to the minimum of the weight of the periods of N/B for all odd N in $(0, B/3)$.

Theorem 6.6 enables the determination of the distance of the AN code by obtaining the periods of N/B for all odd N in $(0, B/3)$. This reduces the computation required for determining distance by a factor of 6. Further reduction in computation can be obtained by use of the "B-orbits" [5] and the number theory properties of B as follows: Let N be an element of $(0, B/3)$, and $q_1 \cdots q_{e(B)}$ be the period of N/B. Then the weight of AN equals the weight of $q_1 q_2 \cdots q_{e(B)}$. To determine the weight of the sequence $q_1 \cdots q_{e(B)}$, we proceed as before with $r_0, r_1, r_2, \ldots$, where $r_i \equiv 2^i N \bmod B$. Applying (6.17) once again, $q_i \neq 0$ if $|r_i|_B$ is in the interval $(B/3, 2B/3]$.

For any $a \in Z_B$, let the sequence $\{a, |2a|_B, |2^2 a|_B, \ldots, |2^{e(B)-1} a|_B\}$ be called the *B-orbit of a*. Then the weight of AN (and also the weight of $q_1 \cdots q_{e(B)}$) equals the number of integers of the B-orbit of N belonging to the interval $(B/3, 2B/3]$. For any x in the B-orbit of N, the weight of Ax is the same as the weight of AN, and therefore we call this the weight of the B-orbit of N. For an arbitrary B, there could be a number of B-orbits.

Consider a prime B. When 2 is primitive in $GF(B)$ there is only one B-orbit, and when -2 is primitive in $GF(B)$ there are two distinct B-orbits. When 2 generates a proper subgroup of $G(B)$, $G(B)$ can be partitioned into cosets of the subgroup generated by 2, and the number of distinct B-orbits equal the number of such cosets. These B-orbits are called *local* B-orbits. The distance of the AN code will then be the minimum of the weights of all these local B-orbits.

For a composite $B = \prod_{i=1}^{t} p_i^{\alpha_i}$, the elements of $G(B)$ can be partitioned once again into local B-orbits. For $N \in Z_B$ but $N \notin G(B)$, the B-orbit of N can be determined as follows. Since N is not relatively prime to B, let N be relatively prime only to a proper divisor B_1 of B. Then $N = kB_2$ for some k and $B_1 B_2 = B$.

$$N \cdot 2^{e(B_1)} = kB_2 \cdot 2^{e(B_1)} = kB_2 \cdot (1 + CB_1) = kB_2 + CkB$$
$$= N + CkB = N \; (\bmod B) \quad \text{for some } c$$

Therefore the B-orbit of N contains a repetition of the sequence $\{N, |2N|_B, \ldots, |2^{e(B_1)-1} N|_B\}$ by $e(B)/e(B_1)$ times. Therefore the B-orbit of N is called the *transferred B-orbit*, and is a concatenation of $e(B)/e(B_1)$

copies of local B_1-orbits. The weight of the B-orbit of N is $e(B)/e(B_1)$ times the weight of the local B_1-orbit of N. All possible transferred B-orbits are searched until these B-orbits exhaust all nonzero elements of Z_B. The minimum distance of the CHP code is then the minimum of the weight of all the local and transferred B-orbits, and therefore we have the following.

THEOREM 6.7

The distance of the cyclic AN code generated by

$$A = (2^{e(B)} - 1)/B$$

is the minimum of the weights of the local and transferred B-orbits.

EXAMPLE

Let $B = 93$. The local B-orbits and their weights are as follows, where an italicized number with underscore indicates that it is in $(B/3, 2B/3]$:

Local B-orbits:	Weight
$F = \{2^j\} = \{1, 2, 4, 8, 16, \underline{32}, 64, \underline{35}, 70, \underline{47}\}$	3
$5F = \{5 \cdot 2^j\} = \{5, 10, 20, \underline{40}, 80, 67, \underline{41}, 82, 71, \underline{49}\}$	3
$7F = \{7 \cdot 2^j\} = \{7, 14, 28, \underline{56}, 19, 38, 76, \underline{59}, 25, \underline{50}\}$	4
$11F = \{11 \cdot 2^j\} = \{11, 22, \underline{44}, 88, 83, 73, \underline{53}, 13, 26, \underline{52}\}$	3
$17F = \{13 \cdot 2^j\} = \{17, \underline{34}, 68, \underline{43}, 86, 79, 65, \underline{37}, 74, \underline{55}\}$	4
$23F = \{23 \cdot 2^j\} = \{23, \underline{46}, 92, 91, 89, 85, 77, \underline{61}, 29, \underline{58}\}$	3

Transferred B-orbits:	
$3F = \{3 \cdot 2^j\} = \{3, 6, 12, 24, \underline{48}, 3, 6, 12, 24, \underline{48}\}$	2
$9F = \{9 \cdot 2^j\} = \{9, 18, \underline{36}, 72, \underline{51}, 9, 18, \underline{36}, 72, \underline{51}\}$	4
$15F = \{15 \cdot 2^j\} = \{15, 30, \underline{60}, 27, \underline{54}, 15, 30, \underline{60}, 27, \underline{54}\}$	4
$21F = \{21 \cdot 2^j\} = \{21, \underline{42}, 84, 75, \underline{57}, 21, \underline{42}, 84, 75, \underline{57}\}$	4
$33F = \{33 \cdot 2^j\} = \{\underline{33}, 66, \underline{39}, 78, 63, \underline{33}, 66, \underline{39}, 78, 63\}$	4
$45F = \{45 \cdot 2^j\} = \{\underline{45}, 90, 87, 81, 69, \underline{45}, 90, 87, 81, 69\}$	2
$31F = \{31 \cdot 2^j\} = \{31, \underline{62}, 31, \underline{62}, 31, \underline{62}, 31, \underline{62}, 31, \underline{62}\}$	5

The weight of a B-orbit is determined by the number of elements of the B-orbit belonging to the middle-third of Z_B, namely the interval (31, 62]. From the weights given above the distance of the code generated by $A = (2^{10} - 1)/93 = 11$ is only 2.

EXAMPLE

Let $B = 39$; the integers belonging to (13, 26] are italicized to facilitate weight count. The generator $A = (2^{12} - 1)/39$, and $d_{min} = 4$.

Local B-orbits: Weight

$F = \{2^j\} = \{1, 2, 4, 8, \underline{16}, 32, \underline{25}, 11, \underline{22}, 5, 10, \underline{20}\}$ 4

$7F = \{2.2^2\} = \{7, \underline{14}, 28, \underline{17}, 34, 29, \underline{19}, 38, 37, 35, 31, \underline{23}\}$ 4

Transferred B-orbits:

$3F = \{3.2^j\} = \{3, 6, 12, \underline{24}, 9, \underline{18}, 36, 33, 27, \underline{15}, 30, \underline{21}\}$ 4

$13F = \{13.2^j\} = \{13, \underline{26}, 13, \underline{26}, 13, \underline{26}, 13, \underline{26}, 13, \underline{26}, 13, \underline{26}\}$ 6

The strategy for better codes

The Barrows–Mandelbaum codes usually have a good distance but suffer from a low information rate. The Brown–Peterson codes usually have a high rate, but a low distance. The determination of distance for codes generated by $A = [2^{e(B)} - 1)/B$ for any odd B involves computation of the local and transferred B-orbits and their weights. The question that can be asked is, How can one choose a B to obtain good rate and good distance at the same time? This question has been attacked by Chien et al. [5] as follows.

Let $B = \prod_{i=1}^{t} p_i^{\alpha_i} (\alpha_i \geq 1)$ for odd primes $p_1, p_2, \ldots, p_t$ which are 2-regular. A 2-regular prime p is one having $e(p)$ different from $e(p^2)$.

It is known that all primes less than 10^6 are 2-regular except 1093 and 3511 [9]. Denoting $e_i = e(p_i)$ for $i = 1, 2, \ldots, t$, we get

$$e(B) = \left\langle \left(\prod_{i=1}^{t} p_i^{\alpha_i - 1} \right), e_1, e_2, \ldots, e_t \right\rangle$$

$$= \left\langle \prod_{i=1}^{t} p_i^{\alpha_i - 1}, \langle e_1, e_2, \ldots, e_t \rangle \right\rangle \tag{6.22}$$

Also we write

$$e(p_1, p_2 \cdots p_t) = \langle e_1, e_2, \ldots, e_t \rangle = K \prod_{i=1}^{t} p_i^{s_i} \tag{6.23}$$

for some K relatively prime to p_i $(i = 1, 2, \ldots, t)$, where the s_i are appropriate integers, possibly zero. Then

$$e(B) = K \prod p_i^{\beta_i} \tag{6.24}$$

where $\beta_i = \max(\alpha_i - 1, s_i)$. Given t distinct primes $p_1, p_2, \ldots, p_t$, the exponents s_i and K are uniquely defined by (6.23). For the same primes we define the saturation product S given by

$$S = S(p_1, p_2, \ldots, p_t) = \prod_{i=1}^{t} p_i^{s_i + 1} \tag{6.25}$$

The importance of S is made clear later but first we observe that $e(S) = K \prod_{i=1}^{t} p_i^{s_i}$. Also for all B that divide S, $e(B) = e(S)$. Since $e(B)$ is the length of the cyclic code generated by $A = (2^{e(B)} - 1)/B$, the length remains the same for all divisors of S.

Noting the expression for rate

$$IR = \log_2 B/e(B) \tag{6.26}$$

$e(B)$ remains constant for all divisors of B, while the numerator is maximum when $B = S$; we can say that the rate is maximized for $B = S$. For $B = \prod_{i=1}^{t} p_i^{\alpha_i}$, a multiple of S, we have $\alpha_i \geq s_i + 1$ and $e(B) = e(S) \cdot B/S$. For increasing α_i, the denominator $e(B)$ grows faster than $\log_2 B$ and therefore IR is once again maximized for $B = S$. Therefore we have the following.

THEOREM 6.8

For the given odd primes $p_1, p_2, \ldots, p_t$, we consider the class of cyclic codes with B of the form $\prod_{i=1}^{t} p_i^{\alpha_i}$, $\alpha_i \geq 1$, and the generator $A = (2^{e(B)} - 1)/B$. In this class of CHP codes the code with the maximum information rate is the one having $B = S(p_1, p_2, \ldots, p_t)$.

Theorem 6.8 is an important contribution, for it provides some guidelines for finding codes with good rate (or reasonable rates) and large distance at the same time. For B's that are multiples of S, the increasing α_i may provide an increasing distance, but the rate will be decreasing. On the other hand, for B's that are divisors of S, as the α_i increase toward S, the rate increases with possibilities of good distance. By the use of computation procedures as suggested in the previous section for values of B close to S, one could obtain optimum codes. Further results in the area relate to techniques for reducing computation required for this search and for a determination of actual distance. For further results in this area and a thorough presentation, the reader is advised to refer to Chien *et al.* [5].

6.4 *AN* CODES FOR COMPOSITE $A = \prod (2^{m_i} - 1)$

A class of AN codes where

$$A = \prod_{i=1}^{r} (2^{m_i} - 1)$$

was first discussed by Kondratyev and Trofimov [10] and later by Monteiro and Rao [11]. Monteiro *et al.* [12] have studied these codes further and obtained the necessary and sufficient conditions for the code to have a distance greater than 2^s for $s < r$. We summarize here the work of these authors [10–12].

Necessary conditions for $d_{min} > 2^s$

Necessary conditions for achieving large d_{min} by general *AN* codes have been given by Kondratyev and Trofimov [10, p. 86] without a formal proof. We comment that what is of interest, once an A is given, is the choice of code modulus M, or equivalently, the code length n.

THEOREM 6.9

Let an *AN* code be formed with

$$A = \prod_{j=1}^{r} (2^{m_j} - 1) \tag{6.27}$$

having the m_j pairwise relatively prime (and ordered); that is,

$$(m_i, m_j) = 1, \quad m_i > m_j > 1 \quad \text{for all} \quad i > j \quad \text{for all}$$
$$j \in \{1, \ldots, r-1\} \tag{6.28}$$

Then necessary conditions for $d_{min} > 2^s$ are that $r > s$, and the code length satisfies

$$n \le n_0 = \min\left(\prod_{j \in I_1} m_j + \prod_{j \in I_2} m_j + \cdots + \prod_{j \in I_s} m_j\right) \tag{6.29}$$

where the minimum is taken over all nonempty disjoint partitions of

$$\{1, \ldots, r\} = \bigcup_{i=1}^{s} I_i.$$

Proof

We first observe that A is a codeword, and on multiplying the terms of (6.27), we find that there are 2^r terms, or perhaps fewer in its NAF, in which case $2^s < d_{min} \le W(A) \le 2^r$; hence, necessarily, $s < r$.

Consider next the integer, for some partition of $\{1, \ldots, r\}$,

$$L = \prod_{i=1}^{s} \left(2^{\prod_{j \in I_i} m_j} - 1\right) \tag{6.30}$$

where the nonempty, nonintersecting sets of integers I_i contain all integers $1, \ldots, r$; that is, $\bigcup_{i=1}^{s} I_i = \{1, \ldots, r\}$. We observe that L is divisible by A and hence is a candidate for a codeword or code modulus. However, $W(L) \leq 2^s$, since L has at most 2^s terms in its NAF expansion, in which case we cannot have L as a codeword. Consequently, since L is divisible by A, $Am \leq L$, while $Am < 2^n$ by definition of n. Partitioning the integers such that the highest exponent of L,

$$\prod_{j \in I_1} m_j + \prod_{j \in I_2} m_j + \cdots + \prod_{j \in I_s} m_j$$

is minimized gives the minimum such L; this exponent is n_0 of (6.29).

Since the next highest power of 2 in L is subtractive we see that the greatest lower bound on n is n_0, i.e., $Am \leq \min L < 2^{n_0}$, from which (6.29) necessarily follows.   Q.E.D.

It is worth commenting that (6.28) is merely for convenience. Indeed if the m_j are not relatively prime, then the same theorem holds except that $\prod m_j$ is replaced by $\text{lcm}(m_j)$ in (6.29). However, the use of m_j that are not relatively prime leads practically to inefficient coding, so it is scarcely considered. We illustrate the theorem numerically for a simple but interesting case.

EXAMPLE

Let it be desired to create an AN code with $d_{\min} > 2^2 = 4$ using the smallest possible A of the form of (6.27). Then $r = 3$ and we consider in the first instance, $m_1 = 2$, $m_2 = 3$, $m_3 = 5$, or

$$A = (2^2 - 1)(2^3 - 1)(2^5 - 1) = 651 = 2^9 + 2^7 + 2^4 - 2^2 - 1$$

where the right side is the NAF, showing that $W(A) = 5$. The set of L's, (6.31), is

$$\begin{aligned}
L_1 &= (2^{2 \cdot 3} - 1)(2^5 - 1) = 2^{2 \cdot 3 + 5} - 2^5 - 2^{2 \cdot 3} + 1 \\
&= 2^{11} - 2^7 + 2^5 + 1 = 1953 = 3 \cdot A, \\
L_2 &= (2^{2 \cdot 5} - 1)(2^3 - 1) = 2^{2 \cdot 5 + 3} - 2^3 - 2^{2 \cdot 5} + 1 \\
&= 2^{13} - 2^{10} - 2^3 + 1 = 7161 = 11 \cdot A, \\
L_3 &= (2^{3 \cdot 5} - 1)(2^2 - 1) = 2^{3 \cdot 5 + 2} - 2^2 - 2^{3 \cdot 5} + 1 \\
&= 2^{17} - 2^{15} - 2^2 + 1 = 98301 = 151 \cdot A
\end{aligned}$$

Thus the minimum L is L_1, from which we see that

$$n_0 = 11 = 2 \cdot 3 + 5 = \min \begin{cases} m_1 m_2 + m_3; & I_1 = \{1, 2\}, & I_2 = \{3\} \\ m_1 m_3 + m_2; & I_1 = \{1, 3\}, & I_2 = \{2\} \\ m_2 m_3 + m_1; & I_1 = \{2, 3\}, & I_2 = \{1\} \end{cases}$$

Consequently we require that $M = Am = 651m < 2048 = 2^{11}$. The maximum m satisfying this inequality is $m = 3$. Choosing this m gives $M = 3A$ and the *AN* code is $\{0, A, 2A\}$ in the ring $Z_{3A} = \{0, 1, \ldots, 3A - 1\}$. In this ring $\bar{A} = 2A$, and $W(A) = W(2A)$ always. Hence $d_{\min} = 5 > 2^2$ for the largest possible n, $n = n_0 = 11$, and thus the conditions of Theorem 6.9 are seen to be sufficient in this case also.

Insufficiency of the necessary conditions

As is used extensively later, in the special case of $s = 2$ the necessary conditions of Theorem 6.9 are known to be sufficient [10, p. 90] to guarantee $d_{\min} > 4 = 2^s$. Consequently the code in the example above does have $d_{\min} > 4$, as has already been seen by inspection. However, we show here by (counter) example that for $s > 2$, the conditions of Theorem 6.9 need not be sufficient for any $n > 0$.

EXAMPLE

Consider an *AN* code with $m_1 = 3$, $m_2 = 4$, $m_3 = 5$, $m_4 = 7$; that is,

$$\begin{aligned} A &= (2^3 - 1)(2^4 - 1)(2^5 - 1)(2^7 - 1) \\ &= 2^{19} - 2^{17} + 2^{14} + 2^{12} - 2^8 - 2^6 + 2^3 + 2^0 \end{aligned}$$

where the NAF on the right shows that $W(A) = 8$. Consequently, any *AN* code generated by A must have $d_{\min} \leq 8$. However, if the conditions of Theorem 6.9 were sufficient, they would allow $d_{\min} > 8 = 2^3$, since $s = 3 < 4 = r$ would be possible for some $n \leq n_0 = 3 \cdot 4 + 5 + 7 = 24$ calculated according to (6.29).

We therefore conclude that, in general, additional constraints on the m_j are necessary to obtain $d_{\min} > 2^s$ for some n satisfying (6.29).

Indeed since we would desire at least one nonzero codeword, meaningful n are $n > \sum\limits_{j=1}^{r} m_j$, the highest exponent in A.

Sufficiency results

Theorem 6.10 of this section offers a simple method for designing a code with $d_{\min} > 2^s$ for any s. Because of their multiresidue implementation we are most interested in those A having the form given in (6.27). However, the primary result of Theorem 6.10 is more general and hence given for arbitrary A.

THEOREM 6.10

Given an AN code with minimum distance $d_{\min}$, length n, and information range m, the $A'N$ code formed with

$$A' = A(2^l - 1) \tag{6.31}$$

has minimum distance $d'_{\min} = 2d_{\min}$, with $m' = m$ for a code length $l + n$ if

$$l \geq n + 1 \tag{6.32}$$

Proof

Let $A \cdot N$ in its NAF be

$$A \cdot N = 2^{j_1} \pm 2^{j_2} \pm \cdots \pm 2^{j_t}$$

for some $t \geq d_{\min}$. Then $A'N$ can be written as

$$A'N = (A \cdot N) \cdot (2^l - 1) = (A \cdot N)2^l - A \cdot N$$

since $l \geq n + 1$ and $j_1 < n$, all the $2t$ terms of $A'N$ are nonadjacent and therefore $A'N$ has weight $2t$.

Thus, given any codeword $A \cdot N$ in the original code, we can show that its weight is doubled when multiplied by $2^l - 1$. We also observe that the information range of the new $A'N$ code is the same as that of the original AN code, from which we conclude that every $A'N$ codeword can be written as $AN(2^l - 1)$. Since every codeword is the original AN code has a weight at least d, the minimum distance of the $A'N$ code is seen to be at least $2d$.   Q.E.D.

To achieve a construction of large distance codes having A of the desired form of (6.27) we iteratively apply (6.31) to an initial A formed according to the known, but little recognized, results for $d_{\min} > 2^2$ of Kondratyev and Trofimov [10]. We thus state the latter here for reference but without proof. We do point out that the code-length definition we use is numerically one greater than that used by Kondratyev and Trofimov in their proof.

THEOREM 6.11

Let an AN code be formed with

$$A = \prod_{j=1}^{r} (2^{m_j} - 1) \tag{6.27}$$

having

$$(m_i, m_j) = 1, \qquad m_i > m_j > 1 \quad \forall i > j \qquad \forall j \in \{1, \ldots, r - 1\} \tag{6.28}$$

Then $d_{\min} > 4$ for all $r \geq 3$ and all positive n satisfying

$$n \leq n_0 = \min\left(\prod_{j \in I_1} m_j + \prod_{j \in I_2} m_j\right) \tag{6.33}$$

where the minimum is taken over all nonempty disjoint partitions of $\{1, \ldots, r\} = I_1 \bigcup I_2$. In other words $n \leq n_0$ is a necessary and sufficient condition for $d_{\min} > 2^s$ for $2 = s < r$. Using any A and n satisfying Theorem 6.11, Theorem 6.10 can be iteratively applied to obtain AN codes with A of the form of (6.27) and having $d_{\min} > 2^s$ for any integer $s \geq 2$.

Theorems 6.10 and 6.11 give a lower bound on $d_{\min}$ for the chosen code. In some cases it may be of interest to also fix an upper bound.

For this we can fix the code modulus $M = Am$ such that a codeword of weight 2^{s+1} occurs within the code for $d_{min} > 2^s$; such a codeword is

$$AN = \prod_{i=1}^{s+1}\left(2^{\prod_{j \in I_i} m_j} - 1\right) \quad (6.34)$$

where the sets I_i are those for which the minimum of $\sum_{i=1}^{s+1}\left(\prod_{j \in I_i} m_j\right)$ occurs when taken over all $s + 1$ nonempty disjoint partitions of $\{1, 2, \ldots, r\} = I_1 \cup \cdots \cup I_{s+1}$. Consequently, we can obtain

$$2^{s+1} \geq d_{min} > 2^s$$

The $s = 1$ situation, where $n_0 = \prod_{j=1}^{r} m_j$, is worth a comment. In this case the necessary conditions of Theorem 1 are seen to be sufficient also. Likewise, in this $s = 1$ case, $d_{min} = 4$ is known, Kondratyev and Trofimov [10, p. 86] (though some $m_j > 3$ must be held, Monteiro [13]). The $s = 1$ code is of some interest since it is the only cyclic code of this class.

Some example codes are given in Table 6.2, along with their information rates, as defined in the next section.

Table 6.2   Information rates for some AN codes with $A = \prod_{j=1}^{r} (2^{m_j} - 1)$ and $d > 2^s$

$A = \prod_{j=1}^{r} (2^{m_j} - 1)$	s	Length	IR
$(2^5 - 1)(2^6 - 1)(2^7 - 1)$	2	37	0.515
$(2^5 - 1)(2^6 - 1)(2^7 - 1)(2^{41} - 1)$	3	78	0.244
$(2^7 - 1)(2^8 - 1)(2^9 - 1)(2^{67} - 1)(2^{137} - 1)$	4	269	0.152
$(2^7 - 1)(2^8 - 1)(2^9 - 1)(2^{11} - 1)(2^{13} - 1)(2^{653} - 1)$	3	1300	0.46
$(2^7 - 1)(2^8 - 1)(2^9 - 1)(2^{11} - 1)(2^{151} - 1)(2^{305} - 1)$	4	605	0.188

Relation to repetition codes—Information rate

The codes treated here can be looked upon somewhat as repetition codes (Massey and Garcia [14, p. 305]) with, however, a change in sign in the repetition. To see this we define an $A'N$ code as one formed via Theorem 6.10; we then note that every $A'N$ codeword is of the form

$$A'N = AN(2^l - 1) = AN2^l - AN \qquad (6.35)$$

In the right hand expression there is, in view of the constraint $l \geq n + 1$, no overlapping of the two terms; that is, the term $AN2^l$ is a repetition of $-AN$ with a sign change. In contrast a true repetition code, called here an $A_R N$ code, would have

$$A_R N = AN(2^l + 1) \qquad (6.36)$$

With the choice $l = n$ the latter yields cyclic $A_R N$ codes when the original AN code is cyclic, since $M_R = A_R m = Am(2^l + 1) = (2^n - 1)(2^n + 1) = 2^{2n} - 1$. This is in contrast to the situation with the $A'N$ codes which are noncyclic.

From the repetitive nature, it is clear, on observing (6.35) and (6.36) that the distances of the $A_R N$ and $A'N$ codes are twice those of the original AN code when subject to $l \geq n + 1$ (the $+1$ in $n + 1$ being to preserve nonadjacency).

In terms of information rate, for the same choice of m the $A'N$ codes are just slightly better than the $A_R N$ codes. For this class of codes (given by Theorem 6.10) Z_M is not a metric space and therefore the minimum distance cannot automatically be related to error control properties. Therefore Monteiro et al. [12] have established error control properties, i.e., the correction of errors of weight $t \leq 2^{s-1}$ by a suitable modification of (6.34). Also they discuss implementation, i.e., encoding and decoding procedures, for this class of codes as well as their corresponding multiresidue codes in some detail.

REFERENCES

1. G. H. Reitwiesner, Binary Arithmetic, *Advances in Comput.* **1**, 232–308 (1960).
2. J. L. Massey, Survey of Residue Coding for Arithmetic Errors, *Intern. Comput. Center Bull.* **3**, (October 1964).
3. S. H. Chang and N. Tsao-Wu, Distance and Structure of Cyclic Arithmetic Codes, *Proc. Internat. Conf. on System Sci., Hawaii, 1968*, pp. 463–466.

4. W. E. Clark and I. J. Liang, Arithmetic Weight For a General Radix Representation of Integers, *IEEE Trans. Inform. Theory*, **IT-19** (November 1973).

5. R. T. Chien, S. J. Hong, and F. P. Preparata, Some Results in the Theory of Arithmetic Codes, *Information and Control* **19**, 246–264 (1971).

6. M. Goto and T. Fukumura, The Distance of Arithmetic Codes, *Mem. Fac. Engrg. Nagoya Univ.* pp. 474–482 (1968).

7. J. T. Barrows, Jr., "A New Method for Constructing Multiple Error Correcting Linear Residue Codes," Rep. R-277. Coordinated Sci. Lab., Univ. of Illinois, Urbana, January 1966.

8. D. Mandelbaum, Arithmetic Codes with Large Distance *IEEE Trans. Information Theory* **IT-13**, 237–242 (April 1967).

9. H. Riesel, Note on the Congruence $a^{p-1} \equiv 1 \pmod{p^2}$, *Math. Comp.* **18**, 149–150 (1964).

10. V. N. Kondratyev and N. N. Trofimov, Error-Correcting Codes with a Peterson Distance Not Less Than Five, *Engrg. Cybernetics* 85–91 (May-June 1969).

11. P. M. Monteiro and T. R. N. Rao, Multi-Residue Codes for Double Error Correction, *Proc. IEEE-TCCA Symp. Comput. Arithmetic, Univ. of Maryland, College Park, May 1972.*

12. P. Monteiro, R. W. Newcomb and T. R. N. Rao, *AN* Codes for Arbitrarily Large Distances, Tech. Rep., Dept. of Elec., Engrg., Univ. of Maryland, College park, October 1972.

13. P. Monteiro, The Theory and Application of *AN* Codes for $A = \prod_{i=1}^{r} (2^{m_i} - 1)$, Ph.D. Thesis, Dept. of Elec. Engrg., Univ. of Maryland, College Park, Maryland, 1972.

14. J. L. Massey and O. N. Garcia, Error Correcting Codes in Computer Arithmetic, Chapter 5, "Advances in Inform. Systems Science" (Ed. J. L. Tow), Vol. 4, 273–326, Plenum Press, New York, 1971.

7 OTHER ARITHMETIC CODES OF INTEREST

In this chapter we study three different types of codes. These are *systematic nonseparate codes, burst error codes, and iterative error codes.*

7.1 SYSTEMATIC NONSEPARATE CODES

We observed in Chapter 3 that AN codes are nonsystematic, and that systematic codes can be further divided into separate and non-separate categories. As separate codes we studied $[N, |N|_b]$ codes and the biresidue code $[N, |N|_{m_1}, |N|_{m_2}]$. An example of a systematic, nonseparate code was given by the $|25N|_{40}$ code (Table 3.1).

Peterson [1] and later Massey [2] observed that for a given AN code, $2^{a-1} < A < 2^a$, a systematic form (or systematic subcode) can be obtained by coding the information N as

$$X = N \cdot 2^a + C_1(N) \qquad (7.1)$$

where the check

$$C_1(N) = |-N \cdot 2^a|_A$$

Clearly the following can be observed: (1) X is a multiple of A and therefore is an AN codeword; (2) the high-order digits of X (in binary form) represent the information N, and the low-order digits represent the check, so the code is thus systematic. As an example, a systematic form of a $3N$ code is obtained by coding N as

$$X = N \cdot 2^2 + C_1(N)$$

where

$$C_1(N) = |-N \cdot 2^2|_3 = |-N|_3 = |2N|_3$$

This code, which represents each N by $X = 4N + |2N|_3$, is given in column 3 of Table 7.1. Massey [2] very correctly observes that each codeword X here is a member of an AN code, but that every AN codeword is not present in this systematic code; he therefore calls it the *systematic subcode*. Further, we observe that this code cannot be closed under ordinary addition, and therefore it is nonlinear. As an example, note that the addition of codewords 00110 to 10010 will result in 11000, which is not present in the systematic subcode. This fact is also clear when two codewords for N_1 and N_2 are added using a conventional binary adder. The carry from the check part propagates into the information and therefore the result will not always be a codeword for $N_1 + N_2$. To eliminate this difficulty Garner [3] introduced a systematic code variety with the information digits on the right and check digits on the

Table 7.1

		Systematic forms of $3N$ codes	
$0 \leq N < 8$	$3N$ code	$X = N \cdot 2^2 + C_1(N)$ information on the left	$y = C_N \cdot m + N$ information on the right
000	00000	00000	00000
001	00011	00110	01001
010	00110	01001	10010
011	01001	01100	00011
100	01100	10010	01100
101	01111	10101	10101
110	10010	11000	00110
111	10101	11110	01111

left; he derived an addition structure for the codewords in such a way that the code is closed under addition and the codeword for $N_1 + N_2$ is represented by the sum. This class of codes is studied later by Rao [4] as $|gAN|_M$ codes. An example of this class of codes is the $|9N|_{24}$ code for $0 \le N < m = 8$ (column 4 of Table 7.1). The $|gAN|_M$ codes are referred to hereafter as gAN codes. The theoretical construction of these codes is considered in the next section.

Construction of a gAN code

Consider an AN code for $0 \le N < m$ such that A and m are relatively prime (m is 2^k for 2's complement adders, and is $2^k - 1$ for 1's complement adders). Then there exists a corresponding systematic gAN code, where

$$M = A \cdot m \tag{7.2}$$

and

$$g \cdot A = C_1 \cdot m + 1 \quad \text{for} \quad 0 < C_1 < A \tag{7.3}$$

gA, the codeword representing 1, can also be written as a pair $(C_1, 1)$. From (7.3) we get $|C_1 m + 1|_A = 0$ or

$$C_1 = |-m^{-1}|_A \tag{7.4}$$

Consequently, the codeword representing $N \in Z_m$ is

$$|gAN|_M = |C_1 \cdot N \cdot m + N|_M = |C_1 N|_A \cdot m + N$$

Denoting

$$C_N = |C_1 \cdot N|_A = |-N \cdot m^{-1}|_A \tag{7.5}$$

we have

$$|gAN|_M = C_N m + N \tag{7.6}$$

which can be represented as a pair (C_N, N).

This leads to a binary representation of the codeword $X = |gAN|_M$ as an n-tuple $(x_{n-1}, x_{n-2}, \ldots, x_k, x_{k-1}, \ldots, x_0)$, where

$$N = \sum_{i=0}^{k-1} x_i 2^i \qquad (7.7)$$

$$C_N = \sum_{j=k}^{n-1} x_j 2^{j-k} \qquad (7.8)$$

This can also be written as $X = |gAN|_M = \sum_{j=0}^{n-1} x_j V_j$, where the V_j are digit values given by

$$V_j = \begin{cases} 2^j & \text{for } j = 0, 1, \ldots, k-1 \\ 2^{j-k} \cdot m & \text{for } j = k, k+1, \ldots, n-1 \end{cases} \qquad (7.9)\dagger$$

Consider two codewords $X = (C_{N_1}m + N_1)$ and $Y = (C_{N_2}m + N_2)$ representing the numbers $N_1, N_2 \in Z_m$, respectively. The sum

$$Z = |X + Y|_M = |(C_{N_1} + C_{N_2}) \cdot m + N_1 + N_2|_M$$

Therefore

$$Z = |(C_{N_1} + C_{N_2} + T) \cdot m + |N_1 + N_2|_m|_M$$

or

$$Z = |C_{N_1} + C_{N_2} + T|_A \cdot m + |N_1 + N_2|_m \qquad (7.10)$$

where

$$T = \begin{cases} 1 & \text{if } N_1 + N_2 \geq m \\ 0 & \text{otherwise} \end{cases}$$

Denoting $N_3 = |N_1 + N_2|_m$, the sum of the information numbers (operands) modulo m, we get

$$Z = |C_{N_1} + C_{N_2} + T|_A \cdot m + N_3 \qquad (7.11)$$

and it follows relatively easily from (7.5) that Z is a codeword representing N_3, and $C_{N_3} = |C_{N_1} + C_{N_2} + T|_A = |-N_3 m^{-1}|_A$. Therefore we have obtained the addition structure for the codewords of the systematic subcode given by Figure 7.1.

† Note that for $m = 2^k$, the digit values given by (7.9) reduce to simply $V_j = 2^j$ for all $j = 0, 1, \ldots, n-1$.

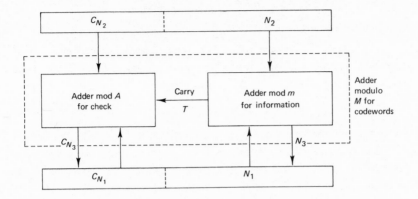

Figure 7.1 Addition structure of a gAN code.

In Figure 7.1, the information parts (the low-order k-bits) of the two codewords are summed using an adder modulo m, and advancing a carry (T) to the check parts if $N_1 + N_2 \geq m$. The check parts C_{N_1}, C_{N_2}, and T are summed by using an adder modulo A to obtain C_{N_3}. The result, obtained as a pair (C_{N_3}, N_3), represents the sum of the codewords modulo M.

Single-error correction

A single error E in the adder modulo M ($M = A \cdot m$) is defined naturally by the form

$$E = \pm V_j$$

where V_j is given by (7.9) and $-V_j$ is synonymous with $M - V_j$. For arbitrary values of A and m these errors are not always of the form $\pm 2^j$. However, if $m = 2^k$ as previously assumed, then the single-error set coincides with $V(M, 1)$, the single modular error set defined in Chapter 2. But for a generalized case the following definitions of weight and distance are appropriate.

DEFINITION 7.1

The systematic weight of $N \in Z_M$, the ring of integers modulo M ($M = A \cdot m$) denoted by $\xi(N)$, is the minimum number of terms in the representation of N or $\overline{N} = M - N$ of the form

$$N \quad \text{or} \quad \overline{N} = a_1 V_{j_1} + a_2 V_{j_2} + \cdots$$

where $a_i = 1$ or -1, and V_{ji} are the digit values as defined in (7.9).

When $m = 2^k$, $V_j = 2^j$ for $j = 0, 1, \ldots, n - 1$, and $\xi(N) = \xi(\overline{N}) = \min\{W(N), W(\overline{N})\}$, and the systematic weight then is the same as the modular weight (W_M) defined in Chapter 2.

EXAMPLE

In $|57N|_{152}$ code, $m = 2^3 = 8$, $A = 19$, and

$$\begin{aligned}
\xi(88) &= \min\{W(88), W(\overline{88})\} \\
&= \min\{W(88), W(64)\} \\
&= 1
\end{aligned}$$

For $x, y, \in Z_M$, let $x - y$ denote the difference in Z_M.

DEFINITION 7.2

The systematic distance between $x, y, \in Z_M$, denoted as $D(x, y)$, is given by $\xi(x - y)$. Then we have

$$D(x, y) = D(y, x) \tag{7.12}$$

$$D(x, y) \geq 0, \qquad \text{equality holds iff} \quad x = y \tag{7.13}$$

For $x, y, w \in Z_M$ ($M = Am$), if $m = 2^k$ and A is of the form that would require no more than two end-around-carries, then from Corollary 2.2 we have

$$D(x, y) + D(y, w) \geq D(x, w) \tag{7.14}$$

If m and A are such that (7.12)–(7.14) hold, then the systematic distance is a metric function. (See, for instance, Birkhoff and MacLane [5]. Also see the note on distances in Section 2.2 of this book.)

DEFINITION 7.3

An error $E \in Z_M$ is said to be a d-fold error or a pattern of d errors iff $\xi(E) = d$.

Consider an AN code for $0 \leq N < 2^k \leq M(A, d)$. This code has a minimum distance of d and therefore is capable of detecting any error of weight $t \leq d - 1$ or correcting any error of weight $t \leq [(d - 1)/2]$. For the same check base A, consider a gAN code where $M = A \cdot m = A \cdot 2^k$ and $2^k \leq M(A, d)$. Clearly each codeword of this code is a multiple of A and is in Z_M. Therefore this code can only be a permutation of the AN code, $0 \leq N < 2^k$. Further, the complement of $x = |gAN|_M \neq 0$ is given by $\tilde{x} = M - x = |gA\bar{N}|_M$, where $\bar{N} = 2^k - N$ and $\tilde{x}$ is also an AN codeword. Thus x and $\tilde{x}$ both have an arithmetic weight greater than or equal to d. Therefore the systematic weight of x is given by

$$\xi(x) = \min\{W(x), W(\tilde{x})\} \geq d$$

Also it can be observed that the minimum systematic weight of the gAN code is no greater than the minimum weight of the AN code. Therefore the minimum systematic weight of the nonzero codewords of the gAN code is exactly d. The minimum systematic distance is the minimum systematic weight of the nonzero codewords. Therefore we have proved the following.

THEOREM 7.1

A gAN code has a minimum systematic distance d, provided

$$m = 2^k \leq M(A, d) \tag{7.15}$$

Consider a gAN code and its addition structure. The adder modulo m (for information) is often a 2's complement or a 1's complement adder. In the latter case, only one end-around-carry connection is used. On the other hand, the addition modulo A may require two or more end-around-carries; this is because when A is of the form 2^c it is of very little use, and when of the form $2^c - 1$ it can provide only error detection, and not error correction. However, it has been previously established (Corollary 2.2), that if M is such that the addition modulo M can be performed with two or fewer end-around-carries, then inequality (7.14) holds for the modular distance in Z_M. Using exactly the same arguments we can state that if A and m are such that additions modulo A and modulo m each require two or fewer end-around-carries, then the systematic distance satisfies (7.14) in Z_M. Using this criterion, we can relate the error-correcting properties of gAN codes to its minimum systematic distance as follows.

THEOREM 7.2

Assume that the systematic distance satisfies (7.14) in Z_M ($M = A \cdot m = A \cdot 2^k$). Then the gAN code can detect any error pattern E of weight d or less [$\xi(E) \le d$] iff the minimum systematic distance is at least $d + 1$. Also this code can correct any error E of weight t or less [$\xi(E) \le t$] iff the minimum systematic distance is at least $2t + 1$.

The relationship of the systematic distance to the error control properties is proved by use of (7.14). This proof is similar to the proof given for Theorem 4.4 so the reader is spared an unneeded and long proof.

Theorems 7.1 and 7.2 can be combined. The error control properties of an AN code can be transferred to its corresponding gAN code provided that (1) addition modulo A requires two or fewer end-around-carries, and (2) $m = 2^k \le M(A, d)$. On the other hand, if A requires three or more end-around-carries, then (7.14) will not hold and the error control properties of the code cannot be established easily. This case is studied by Trehan [6] and by Varanasi and Rao [7]. Their results can be summarized as follows.

THEOREM 7.3 [7]

A gAN code with $m = 2^k$ is t-error correcting iff

$$m < M(A, 3) \qquad \text{for} \quad t = 1$$
$$m < M(A, 3)/2 \qquad \text{for} \quad t > 1$$

The proof of this theorem is rather long and is omitted here, but the interested reader is advised to refer [7].

Error-correcting adder organization

The addition structure of gAN codes was derived earlier. The modulo M adder for codewords is obtained by a combination of two adders, one adder modulo m for the information, and the other modulo A for the checker (see Figure 7.1). This organization has a close resemblance to the separate adder and checker of $[N, |N|_b]$ codes (see Chapter 3) but has two significant differences: (1) there is a carry T from the information adder to the checker for the systematic subcodes; and (2) the gAN code can provide error correction if A is properly chosen and k is limited to the necessary size.

To simplify the error correction procedure, we assume here that $m = 2^k$, and therefore a binary 2's complement adder of k stages is employed for information parts. If $A = 2^c - 1$, then the check addition requires a c-stage 1's complement adder. Such a code can provide only error detection. If the code is to correct single errors (i.e., all errors of the type $+2^j$ or $M - 2^j$ for $j = 0, 1, \ldots, n - 1$), then A may be a prime with 2 or -2 as primitive element of the field of integers modulo A. For instance, if $A = 29$, the check adder of five stages will return two end-around-carries, since $32 = 2 + 1 \pmod{29}$. If $A = 53$, the check adder will be of six stages, and the number of end-around-carries will be three, since $53 = 8 + 2 + 1 \pmod{53}$, and so on. In order for Theorem 7.2 to hold, the selection of A must be such that there are no more than two end-around-carries. Table 7.2 lists some A's which satisfy that condition and provide single-error correction for all values of $k < k_{max}$. [Note that $2^k \leq M_2(A, 3)$.] Also c denotes the number of check bits given by the smallest integer greater than or equal to $\log_2 A$.

The appropriate check base A to be selected from Table 7.2 depends on the information size k. Figure 7.2 describes an organization of an error-correcting adder using nine information bits ($k = 9$) and a check

Table 7.2[a]

A	$M_2(A, 3)$	k_{max}	c
19	$(2^i + 1)/19$	4	5
29	$(2^{14} + 1)/29$	9	5
47	$(2^{23} - 1)/47$	17	6
61	$(2^{30} + 1)/61$	24	6
121	$(2^{55} + 1)/121$	48	7

[a] $M_2(A, 3)$ is obtained by use of Theorem 4.10 on binary AN codes for all A in the table, except for $A = 121$, for which Theorem 4.6 is used.

base $A = 29$ ($c = 5$). The codewords $X = (C_{N1} \cdot m + N_1)$ and $Y = (C_{N2} m + N_2)$, representing the numbers N_1 and N_2, respectively, are each of 14 bits. The codeword X is assumed to be available initially in the accumulator register. The addend N_2 is applied to the encoder, which generates $C_{N2} = |-N_2 2^{-9}|_{29} = |3 \cdot N_2|_{29}$ and stores it in the check part of the addend register. The adder modulo M (of the codewords) consists of the two adders, the 2's complement adder (of nine stages) for the information, and the mod 29 adder (of five stages) for the check. The decoder receives the sum Z' (possibly erroneous) from the accumulator and generates the syndrome

$$S = |Z'|_{29} = ||X + Y + E|_M|_{29} = |E|_{29}$$

Since for all errors $E = 2^j$ or $M - 2^j$, its syndrome $|E|_{29}$ is distinct, the syndrome decoder can generate the position (j) and polarity ($+$ or $-$) of the error. The error corrector shown applies this correction to Z' to obtain the correct output Z. The residue generator, syndrome decoder, and error corrector schemes discussed earlier (Chapter 4) can also be used for gAN codes.

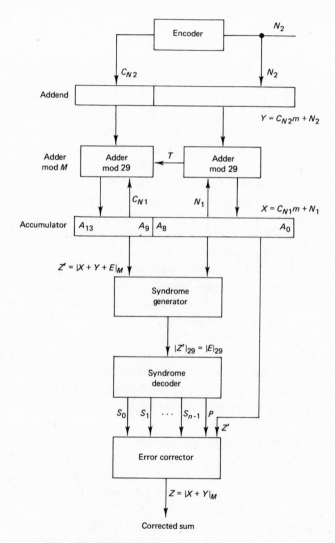

Figure 7.2 Error correction schematic.

7.2   BURST-ERROR-CORRECTING CODES

Although the initial motivation in our effort to develop arithmetic codes was to improve the reliability of arithmetic computation, it is significant that these codes also have some burst-error-detecting and correcting capabilities. We define a burst error E of length b as an integer $B2^j$ (within the appropriate range) where B is an odd integer such that $2^{b-1} < B \leq 2^b - 1$ and is called a burst error pattern.

Henderson [8] was the first to suggest the detection of burst errors of length b by "casting out $(2^{b+1} - 1)$'s." The obvious advantage in such an implementation is due to the facility of formation and addition of the remainders in a radix 2 system. Henderson also suggested the use of a prime p for which 2 is not a primitive element of $GF(p)$ and such that each of the cosets of the subgroup generated by 2 in the multiplicative group of $GF(p)$ includes one (and only one) of the burst error patterns of length b, other than those of the form 2^i, which are contained in the subgroup. Consider, for example, $p = 89$ for which 2 forms a subgroup of order 11 in the multiplicative group of integers modulo 89. The possible error patterns B of length 2 or less have magnitude 1, 2, and 3, but the burst errors they generate are of the form $\pm 2^i$ or $\pm 3 \cdot 2^i$. These cosets may be found for the subgroup generated by 2^i such that -2^i, $+3^i$, and -3^i for $0 \leq i < 11$ are in each of them. Therefore, it would be possible to identify a burst error uniquely by its position in a standard decoding array.

Unfortunately Henderson did not specify whether he intended the coding of the integers to be in a nonseparate (AN) or multiresidue form. From his mention of "casting out p's" the reader may conclude that he had in mind a form of residue coding for which the residue checkers are assumed to be faultless.

Stein [9] expanded on the work of Henderson by noting that if B_i and B_j are any two allowable burst error patterns from a given set of burst error patterns they are in different cosets if the modulo p product of B_i and B_j^{-1} is not in the subgroup of the multiplicative group modulo p generated by 2. Therefore choosing a p for which 2 is not a primitive element and considering all the burst error patterns of a given length, possibly of both signs, we could determine whether or not each pattern is in a different coset. We note that if the order of the subgroup generated by 2 is even, then -1 is in the subgroup, and only positive patterns have to be tested against the sufficient condition stated. Also notice that if we deal with finite ring arithmetic and consider also as allowable burst errors those of the form $B2^j$ for a nonnegative integer j possibly as large as one less than the order of the group, then the condition given is necessary as well as sufficient for correction, provided, of course, that the codeword length $n \le (p-1)/2^b$ for burst error patterns of length b or less. This illustrates the concept of *circular* burst errors in which the error digits "spilled over" the high-order positions of the codeword may affect some of the lower order positions; for example, in a modulo 7 system three circular bursts of length 2 are possible, namely 011, 110, and 101. This, however, is not a standard way of treating burst errors in general.

Independently Chien [10] reached more general conclusions than those of Stein by considering $G(A)$, the multiplicative group in Z_A (see Section 1.2 for definitions). It is known that the order of this group is $\phi(A)$ where ϕ is the Euler ϕ function. If A is a prime, then $\phi(A) = A - 1$ and the group consists of all integers $1, 2, \ldots, (A - 1)$. If $A = p^\alpha$ for a prime p, then $\phi(A) = p^{\alpha-1}\phi(p)$. Furthermore, if A is factored into relatively prime factors B and C, then $\phi(A) = \phi(B)\phi(C)$. The previous information should allow the computation of the Euler ϕ function of any integer by recursion if necessary.

Chien also considered cases where A is not prime but either a power of a prime or a composite integer. Because of the importance of the results obtained by Chien we study them here in some detail, and try to emphasize the similarities and differences between them and other approaches taken. One of the latter is the consideration of these codes as "linear" codes imbedded in the *infinite* ring of integers. If we impose the constraint that N_1 and N_2, two information integers to be encoded, add to less than m, then $A(N_1 + N_2) = AN_1 + AN_2$ and therefore the code is linear. This approach vividly contrasts with considerations in which the AN codes are thought of as ideals in a *finite* ring. Also of importance in burst error correction is the fact that, for an n-bit code and a burst of length $b > 1$, only $(n - b)$ syndromes are required to correct the error. If the burst error is $B \cdot 2^j$ for $j > (n - b)$ in the n-bit codeword, some of the b bits would fall outside the n bits of the length of the codeword, and if those overflowing bits are neglected, the burst would correspond to an error pattern of shorter length. The number of possible burst errors of a given length $b > 1$ are then, under these considerations, $2 \cdot (2^{b-2})(n - b + 1)$, where the factor 2 is due to the possible negative or positive sign of the errors. If we considered these errors distinct (in finite ring arithmetic, a negative burst error of a given length could of course be equivalent to another burst error of the same or different length), the total number of burst errors of length b or less would be $2\left(n + \sum_{i=2}^{b} 2^{i-2}(n - i + 1)\right)$.

This approach has not been taken in the past, and both Stein and Chien consider the burst errors distinct from those errors in which the overflow bits are discarded. For example, in a 10-bit $41N$ code they consider that the error 3×2^9 or 11000000000 is different from the error 2^9. It is in this sense that Chien defines an AN code of length n as a *compact* code if the integers of $G(A)$ correspond exactly to the residues of $B2^j$ modulo A for $j = 0, 1, \ldots, n - 1$ and $B \in \{\pm 1, \pm 3, \ldots, \pm (2^b - 1)\}$. Note that under this concept of a compact code 2^b values for B are possibly multiplied by n powers of 2 to yield $n \cdot 2^b$ possible errors. Consider now the group $G(A)$ and let H denote the subgroup of $G(A)$ generated by 2. If $G(A)$ is an n-bit compact code that corrects all bursts of length b or less in the sense of Chien, then its order $\phi(A)$ must equal $n \cdot 2^b$, that is

$$n = \phi(A)/2^b$$

Also since there are 2^b error patterns, the number of error patterns corrected per coset of a standard array must be the same, and therefore divide 2^b, i.e., such a number must be a power of 2. We will denote this number by j, and also denote a compact (in the sense of Chien) code capable of correcting all bursts of length b or less by C_{bj}. If A is a prime power, then $G(A)$ is a cyclic group, and it can be shown that there will be C_{b1} codes if each of the elements of the set $B = \{\pm 1, \pm 3, \ldots, \pm(2^b - 1)\}$ is in a distinct coset of H, and C_{b2} codes if the elements of $\{1, 3, \ldots, (2^b - 1)\}$ are in different cosets of H. For A a prime, Chien shows the nonexistence of C_{21} and C_{24} codes, and the possibility of C_{22} codes only when $A = 24k + 17$.

7.3 ITERATIVE ERRORS

A functional block (or a logic circuit) is utilized a number of times during the course of an arithmetic operation. For instance a (parallel) byte adder stage is used iteratively in a serial manner a number of times for an addition operation. Such an adder is called a *serial–parallel adder* and is sketched in Figure 7.3. Two n-bit operands A and B are initially contained in registers **A** and **B**. Each operand is divided into r bytes (sometimes also called blocks) of length c bits each. Here $n = rc$. The two low-order bytes $A^{(0)}$ and $B^{(0)}$ are added parallel in the byte

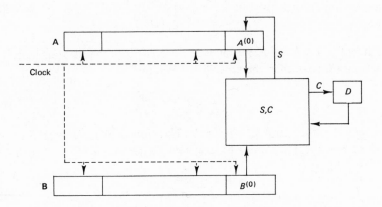

Figure 7.3 Serial–Parallel adder.

adder which generates a sum S (one byte) and a carry C (one bit). Initially D is set to zero. The transfers for the operation can be written as

$$S(A^{(0)}, B^{(0)}, D) \rightarrow A^{(0)}$$
$$C(A^{(0)}, B^{(0)}, D) \rightarrow D$$

The registers **A**, **B** are cyclically shifted by c places to bring the next bytes in position for addition. This (partial) operation is repeated r times. The sum is formed in the register **A** while **B** is left with the original operand B.

Another example of an iterative use of a logic circuit is in high-speed multiplication operations. The multiplier is divided into r bytes of c bits each and each time a partial product (of multiplicand and a byte of the multiplier) is generated using a byte multiplier logic block. These partial products are appropriately shifted and added together in the multiple-operand adder [10, 11].

Errors generated by such multiple-use (iterative-use) blocks are termed *iterative errors*. Assume that an output signal of this logic block is stuck at 1, at the jth stage permanently. Then the error for the first iteration is 2^j or 0, depending on whether the correct signal is a 0 or a 1. Similarly the error in the second iteration is 2^{j+c} or 0, and the error E for the r block operation is then

$$E = +2^j \sum_{i=0}^{r-1} e_i 2^{ci}$$

where $0 \leq j < c$ and $e_i = 0$ or 1. If the error is in the lth stage and is a stuck-at-0 type, then the error E is given by

$$E = -2^l \sum_{i=0}^{r-1} e_i 2^{ci}$$

where $0 \leq l < c$ and $e_i = 0$ or 1. Therefore, in iterative processors using block length c and r blocks, a single iterative error is of the form

$$E = \pm 2^j \sum_{i=0}^{r-1} e_i 2^{ci} \tag{7.16}$$

where the error position j is $0 \leq j < c$, polarity $+$ denotes stuck-at-1 fault, $-$ denotes stuck-at-0 fault, and $e_i = 0$ or 1 for all i.

To detect iterative single errors, a separate checker using base $b = 2^c - 1$ is conveniently employed. (Note that c here coincides with the block length.) Then the syndrome

$$S = |E|_b = \left| \pm 2^j \sum_{i=0}^{r-1} e_i 2^{ci} \right|_{2^c-1}$$

Noting that $(2^c - 1, 2^j) = 1$ and $|2^{ci}|_{2^c-1} = 1$ for all i, we obtain

$$S \neq 0 \quad \text{iff} \quad \left| \sum_{i=0}^{r-1} e_i \right|_b \neq 0 \tag{7.17}$$

It is trivially seen that the above is satisfied for any e_i, if $r < b$. Therefore we obtain the following.

THEOREM 7.4†

For an iterative processor operation using r bytes of length c bits each, all single iterative errors are detected by a separate checker with base $b = 2^c - 1$, provided

$$r < b$$

The above provides an upper limit r_{max} for the number of bytes and therefore limits the length n of the operands (Table 7.3). It should be noted that Theorem 7.4 also holds for any b, provided b is a divisor of

Table 7.3

c	$r_{max} = b - 1$	n_{max}
2	2	4
3	6	18
4	14	56
5	30	150

† A similar theorem for AN codes is stated by Chien and Hong [12] for detecting errors in high-speed multipliers. Avizienis [13] had previously obtained a similar result for serial byte adders.

$2^c - 1$. But when b is a proper divisor of $2^c - 1$, the error-checking logic is not as simple, and also r_{max} and n_{max} are appropriately smaller.

For high-speed multiplication, the byte size c is very critical. For increasing values of c the speed of multiplication increases proportionately, while the cost of the partial product generator increases very rapidly. Therefore c is usually 2 or 3 and rarely greater than 4. For small c, the operand size is too small. Therefore a check base $b = 2^{cl} - 1$ for some $l > 1$ is considered by Varanasi and Rao [14]. For such a base it can be proved that $r_{max} = l(2^c - 1) - 1$, which means much larger operand sizes, as indicated in Table 7.4.

Table 7.4 n_{max}

c \ l	2	3	4	5
2	10	16	22	28
3	39	60	81	102
4	116	176	236	296

Chien and Hong [12] have also derived classes of AN codes capable of correcting single iterative errors or detecting double iterative errors. For correcting single iterative errors, A is chosen to detect the polarity, the error position j, and the error distribution number e_i. This theory is detailed in their paper and the interested reader is advised to refer to their work [12].

REFERENCES

1. W. W. Peterson, "Error-Correcting Codes." MIT Press, Cambridge, Massachusetts 1961.
2. J. L. Massey, Survey of Residue Coding for Arithmetic Errors, *Intern. Comput. Center Bull.* 3, (October 1964).

3. H. L. Garner, Error Codes for Arithmetic Operations, *IEEE Trans. Electron. Comput.* **EC-15**, 763–770 (October 1966).

4. T. R. N. Rao, Error Correction in Adders Using Systematic Subcodes, *IEEE Trans. Comput.* **C-21**, 254–259 (March 1972).

5. G. Birkhoff and S. MacLane, "A Survey of Modern Algebra," 3rd ed. Macmillan, New York, 1965.

6. A. K. Trehan, Residue Codes and Their Application to Arithmetic Processors, Ph.D. Thesis, Dept. of Elec. Engrg., University of Maryland, College Park, Maryland, 1971.

7. M. R. Varanasi and T. R. N. Rao, A note on systematic Non-separate Codes, *Proc. Allerton Conf. on Inform. System Sciences*, **10**, Univ. of Illinois, Urbana, Illinois, October 1972.

8. D. S. Henderson, Residue Class Error Checking Codes, *Proc. Nat. Meeting Assoc. Comput. Mach., Los Angeles, California, 1961*. Abstract appears in *Comm. ACM* **4**, 307 (July 1961).

9. J. J. Stein, Prime Residue Error Correcting Codes, *IEEE Trans. Information Theory* **IT-10** (April 1964).

10. R. T. Chien, On Linear Residue Codes for Burst Error Correction, *IEEE Trans. Information Theory* **IT-10**, 127–133 (April 1964).

11. O. L. MacSorley, High Speed Arithmetic in Binary Computers, *Proc. IRE* **49**, 67–91 (January 1961).

12. R. T. Chien and S. J. Hong, Error Correction in High-Speed Arithmetic, *IEEE Trans. Comput.* **C-21**, (May 1972).

13. A. Avizienis, Arithmetic Error Codes: Cost and Effectiveness Studies for Application in Digital Systems, *IEEE Trans. Comput.* **C-20**, 1322–1330 (November 1971).

14. M. R. Varanasi and T. R. N. Rao, A Note on the Detection of Iterative Arithmetic Errors, to be published.

8
RECENT RESULTS ON ARITHMETIC CODES AND THEIR APPLICATIONS

In this chapter we start with an analogy between the polynomial cyclic codes and the cyclic AN codes. This leads us to an interesting conjecture on the minimum distance bound for AN cyclic codes analogous to the Bose–Chaudhury–Hocquenghem bound (BCH bound) for the polynomial codes. Presented in Section 8.3 is a special case of an upper bound for minimum distance when the length of the AN codes is of the form $n = cn_1$. In Section 8.4 a class of codes known as *majority decodable arithmetic codes* is discussed. In Section 8.5 we consider a recent illustration of a separate residue checker for arithmetic and logical operations of a processor. We conclude in Section 8.5 with a brief discussion of arithmetic coding applications such as in the JPL STAR computer.

8.1 POLYNOMIAL CYCLIC CODES AND CYCLIC AN CODES

In communication, the polynomial cyclic codes are an important and well-established class of codes. Here we present a brief discussion of their structure, for the specific purpose of drawing some very interesting analogies between this class and the cyclic AN codes. For a detailed and complete knowledge of these codes the reader is advised to refer to Peterson and Weldon [1] or Berlekamp [2].

A polynomial cyclic code (of some suitable length n) is defined as follows. A code vector $V = (a_0, \ldots, a_{n-1})$, where a_i $(i = 0, \ldots, n - 1)$ is an element of a finite field $GF(q)$, is also considered as a polynomial $V(x) = a_0 + a_1 x + \cdots + a_{n-1} x^{n-1}$. Each code polynomial is a multiple of a generator polynomial $g(x) = g_0 + g_1 x + \cdots + x^m$ which is a divisor of $x^n - 1$. Also any $V(x)$ of degree less than n and which is a multiple of $g(x)$ is a code polynomial.

Consequently, the code has the structure of an ideal generated by $g(x)$ in the algebra (also the ring) of polynomials modulo $x^n - 1$.

Let $V(x)$ be a code polynomial transmitted over a channel and let $V'(x)$ be the polynomial received. Then $V'(x) = V(x) + E(x)$, where $E(x)$ is called the error polynomial. For a single error in the jth position $(0 \leq j < n)$, $V(x)$ and $V'(x)$ differ only in the jth position and $E(x)$ then equals x^j. The *Hamming weight* of a polynomial $V(x)$ (or vector V) is defined as the number of nonzero coefficients in $V(x)$. The *Hamming distance* between two polynomials $U(x)$ and $V(x)$ is the Hamming weight of their difference $U(x) - V(x)$ and is also given by the number of positions by which the coefficients of $U(x)$ and $V(x)$ differ. The minimum Hamming distance of a polynomial code is the minimum of the distances between all distinct pairs of code polynomials. The Hamming distance does satisfy the positive-definite, symmetry, and triangle inequality properties and therefore is a metric function. By virtue of this, the Hamming distance of a polynomial code plays a role similar to that of the arithmetic distance (or modular distance) in AN codes.

As an illustration, we consider an example of a binary Hamming cyclic code of minimum Hamming distance 3 and of length $n = 15$. This code is generated by an irreducible polynomial $g(x) = x^4 + x + 1$ over $GF(2)$. Every polynomial of degree less than 15 is a code polynomial iff it is a multiple of $g(x)$. Note that $g(x)$ divides $x^{15} - 1$, and $h(x) = (x^{15} - 1)/g(x)$ is called the *parity check* polynomial. The code is linear in the sense that the sum (and difference) of two code polynomials is a code polynomial. The minimum Hamming distance for this code is 3 and is therefore single-error correcting or double-error detecting. Further it can be observed that a root α of $g(x)$ is a primitive element of $GF(2^4)$, an extension field of $GF(2)$. That is, α generates all the nonzero elements of $GF(16)$, and 15 is the order of α ($\alpha^{15} = 1$). We define the syndrome of a received polynomial $V'(x)$ as $S(V'(x)) = r(x)$, which is the residue of smallest degree modulo $g(x)$. Clearly, $S(V'(x)) = S(V(x) + E(x)) = S(E(x))$. For the code thus described, the residues of all single-error polynomials [$S(x^j)$, $0 \leq j < n$] are all distinct and therefore the code provides single-error correction. .

The Hamming distance 3 polynomial code is analogous to the cyclic *AN* code (BP code) generated by a prime A having -2 as primitive in $GF(A)$. [Note that if 2 is primitive in $GF(A)$, the cyclic *AN* code has distance 2 only.] Here the analogy is between a primitive polynomial generator $g(x)$ and the prime generator A having -2 as the primitive element in $GF(A)$. The dual code of the Hamming code is the maximal length sequence code, and here the generator polynomial of one is the parity check polynomial of its dual code. The dual code of the BP code is the BM code whose generator $A = (2^{e(B)} - 1)/B$, where B is a prime having -2 as the primitive element. Clearly, the generator and range

Table 8.1 Analogy between cyclic codes

Binary cyclic polynomial codes	Binary cyclic *AN* codes
1. Generator $g(x)$	Generator A
2. Parity check polynomial $h(x) = (x^n - 1)/g(x)$	Information range $m = (2^n - 1)/A$
3. Algebra of polynomials modulo $x^n - 1$	Ring of integers modulo $2^n - 1$
4. Primitive polynomial $g(x)$	Prime A having -2, but not 2, primitive in $GF(A)$
5. Hamming distance 3 code	Brown–Peterson distance 3 code
6. Maximal length sequence code (dual of Hamming code) $g(x) = (x^n - 1)/p(x)$ $p(x)$ is primitive	Barrows–Mandelbaum code (dual of BP code) $A = (2^{e(B)} - 1)/B$ B is prime with -2, but not 2, primitive in $GF(A)$

m of the BP code are the range and generator, respectively, of the BM code. Thus the analogy between the polynomial cyclic codes and the cyclic AN codes is strong and appears to get stronger. As a further evidence of this we discuss in Section 8.3 an interesting conjecture of Chein and Hong [3] on the minimum distance bound of cyclic AN codes. The Chein–Hong bound is analogous to the BCH bound (the distance bound of the BCH codes).

In Table 8.1 we summarize the above-mentioned analogies between the respective quantities and symbols of the binary code classes.

8.2   BCH CODES AND BCH BOUND

BCH codes are polynomial cyclic codes which are, in a general sense, extensions to Hamming codes and provide arbitrarily large distances [1, 2]. The generator polynomial $g(x)$ of a t-error-correcting binary BCH code of length n is given by

$$g(x) = \langle P(x), m_3(x), \ldots, m_{2t-1}(x) \rangle$$

where $P(x)$ is a primitive polynomial of degree m, its root α is of order $n = 2^m - 1$, and $m_i(x)$ $(i = 3, 5, \ldots, 2t - 1)$ is the minimum polynomial of α^i. Note that when $t = 1$, it is a distance 3 Hamming code. Inclusion of each successive factor, $m_3(x)$, $m_5(x)$, etc., increases the minimum distance by 2 (and increases the number of errors to be corrected by one more). The roots of $P(x)$ and $m_i(x)$ $(i = 3, 5, \ldots, 2t - 1)$ are also roots of $g(x)$ and therefore are roots of $x^n - 1$. A root of $P(x)$ namely α, can also be considered as a primitive nth root of unity, and the other roots of $g(x)$ can be expressed as powers of α.† It can be shown that $\alpha, \alpha^2, \alpha^3, \ldots, \alpha^{2t}$ consecutive powers of α are among the roots of $g(x)$, and therefore the code is of minimum distance at least $2t + 1$ and hence can correct t or fewer errors. This fact has been established as the Bose–Chaudhury theorems and the minimum distance (lower)

† Note here that α is an element of $GF(2^m)$, a finite field.

bound given by the theorem is commonly called the BCH bound. Further, the BCH code provides a direct approach to the derivation of codes capable of correcting an arbitrary number of errors. A class of cyclic AN codes analogous to BCH codes could therefore be of great interest and use, but unfortunately there has been only limited success. It is set forth in the form of an interesting conjecture [3] in the following section.

8.3   ON BOUNDS FOR $d_{\min}$ OF CYCLIC AN CODES

Consider the roots of the polynomial $x^n - 1$ in the complex field. These roots are distributed on a circle with the center as origin and radius equal to 1. A primitive nth root of unity here is $\alpha = e^{2\pi i/n}$ and the other roots are $\alpha^2, \alpha^3, \ldots, \alpha^{n-1}$, and $\alpha^n = 1$. Then we have

$$x^n - 1 = \prod_{i=1}^{n} (x - \alpha^i) \tag{8.1}$$

A *cyclotomic* polynomial $Q_h(x)$ is defined as

$$Q_h(x) = \prod_{(i,\, h) = 1} (x - \beta^i) \tag{8.2}$$

where β is a primitive hth root of unity. It is well established [2] that a cyclotomic polynomial has only integer coefficients and satisfies

$$x^n - 1 = \prod_{h \mid n} Q_h(x) \tag{8.3}$$

and

$$Q_h(x) = \prod_{i \in I_h} (x - \alpha^i) \tag{8.4}$$

where

$$I_h = \{i \mid 1 \le i \le n,\, (i, n) = n/h\} \tag{8.5}$$

From (8.3) we can write as examples

$$x - 1 = Q_1(x)$$

$$x^2 - 1 = Q_1(x) \cdot Q_2(x) \qquad \text{and} \qquad \alpha_2(x) = \frac{x^2 - 1}{x - 1} = x + 1$$

$$x^3 - 1 = Q_1(x)Q_3(x) \qquad \text{and} \qquad Q_3(x) = \frac{x^3 - 1}{x - 1}$$

$$x^4 - 1 = Q_1(x) \cdot Q_2(x)Q_4(x) \qquad \text{and} \qquad Q_4(x) = \frac{x^4 - 1}{x^2 - 1}$$

and

$$x^5 - 1 = Q_i(x)Q_5(x) \qquad \text{and} \qquad Q_5(x) = \frac{x^5 - 1}{x - 1}$$

Let α be the primitive fifteenth root of unity. Then from (8.3) we have

$$(x^{15} - 1) = Q_1(x)Q_3(x)Q_5(x)Q_{15}(x)$$

and from (8.4) we have that

$Q_1(x)$ has roots $\alpha^{15} = 1$
$Q_3(x)$ has roots $\alpha^5, \; \alpha^{10}$
$Q_5(x)$ has roots $\alpha^3, \; \alpha^6, \; \alpha^9, \; \alpha^{12}$

and

$Q_{15}(x)$ has roots $\alpha, \; \alpha^2, \; \alpha^4, \; \alpha^7, \; \alpha^8, \; \alpha^{11}, \; \alpha^{13}, \; \alpha^{14}$

This root approach has been successfully used in BCH codes to obtain all the roots of the generator polynomial and to count the successive powers of α included among them. A similar approach can be used for the cyclic AN codes as follows. We replace x with 2 in the above and write

$$A_h = Q_h(2) = \prod_{i \in I_h} (2 - \alpha^i) \tag{8.6}$$

which are integer factors of $2^n - 1$. Each of these integer factors may not be a prime but has a definite relation to the nth roots of unity given by (8.5). We can write for illustration

$$A_1 = 1, \qquad A_2 = 3, \qquad A_3 = 7, \qquad A_4 = (2^4 - 1)/(2^2 - 1) = 5$$
$$A_5 = 2^5 - 1 = 31, \qquad A_6 = (2^6 - 1)/((2^3 - 1)(2^2 - 1)) = 3$$
$$A_{10} = (2^{10} - 1)/(A_5 A_2) = 11$$

For a given A that is a divisor of $2^n - 1$, A is written as a product of the A_h's, where h divides n, and then the number of consecutive powers of $\alpha = e^{2\pi i/n}$ are counted among the roots of the A_h's. However, factorization of A will not be unique in general and there will be ambiguity in their selection due to the fact that the A_i's are not always pairwise prime. To resolve this, Chien and Hong define $A_i(i|n)$ as a strong (or positive) factor of A if this A_i appears on all possible expressions of A as products of distinct $A_i(i|n)$. Clearly, an A_i dividing A is a strong factor if A_i is relatively prime to all other A_i dividing A. Also, A_i and A_j $(i \neq j)$, $(A_i, A_j) \neq 1$, are both strong factors only if $A_i \cdot A_j$ divides A and $(A_i A_j, A_k) = 1$ for all other A_k dividing A. As examples, when $n = 18$, the A_i (for $i|n$) and their roots are as follows:

$$A_1 = 1$$

$A_2 = 3 \qquad$ has root $\qquad \alpha^9$

$A_3 = 7 \qquad$ has roots $\qquad \alpha^6, \ \alpha^{12}$

$A_6 = 3 \qquad$ has roots $\qquad \alpha^3, \ \alpha^{15}$

$A_9 = 73 \qquad$ has roots $\ \alpha^2, \ \alpha^4, \ \alpha^8, \ \alpha^{10}, \ \alpha^{14}, \ \alpha^{16}$

$A_{18} = 3 \cdot 19 \qquad$ has roots $\ \alpha, \ \alpha^5, \ \alpha^7, \ \alpha^{11}, \ \alpha^{13}, \ \alpha^{17}$

If $A = 7 \cdot 73$, then there is no ambiguity and it has two strong factors. If $A = 3 \cdot 19 \cdot 73$, then the possible expressions for A are $A = A_2 \cdot A_9 \cdot 19 = A_6 \cdot A_9 \cdot 19 = A_9 \cdot A_{18}$. And clearly A_9 is the only strong factor. On the other hand, for $A = 3^3 \cdot 19 \cdot 73$, we have $A = A_2 \cdot A_6 \cdot A_9 \cdot A_{18}$ as the only possible way of factoring, where all four factors are strong. Note here that we take each A_i only once in this factorization. After this factorization and selection of strong factors we then look for the roots that are consecutive powers of α covered by these factors as follows:

$A = 3 \cdot 19 \cdot 73 \to$ one strong factor $\{73\} \to$ one root only

$A = 3^3 \cdot 19 \cdot 73 \to$ four strong factors $\{3, 3, 73, 3 \cdot 19\}$

$\qquad = A_2 \cdot A_6 \cdot A_9 \cdot A_{18}$

$\qquad \to \{\alpha^9, \quad \alpha^3, \quad \alpha^{15}, \quad \alpha^2, \quad \alpha^4, \quad \alpha^8, \quad \alpha^{10}, \quad \alpha^{14}, \quad \alpha^{16}, \quad \alpha, \quad \alpha^5,$
$\qquad \quad \alpha^7, \quad \alpha^{11}, \quad \alpha^{13}, \quad \alpha^{17}\}$

$\qquad \to \{\alpha, \quad \alpha^2, \quad \alpha^3, \quad \alpha^4, \quad \alpha^5\}$ five consecutive powers as roots

$A = 3^3 \cdot 19 \cdot 7 \to$ four strong factors $\{3, 3, 3 \cdot 19, 7\}$

$\qquad = A_2 \cdot A_6 \cdot A_{18} \cdot A_3$

$\qquad \to \{\alpha^9, \quad \alpha^3, \quad \alpha^{15}, \quad \alpha, \quad \alpha^5, \quad \alpha^7, \quad \alpha^{11}, \quad \alpha^{13}, \quad \alpha^{17}, \quad \alpha^6, \quad \alpha^{12}\}$

$\qquad \to \{\alpha^5, \quad \alpha^6, \quad \alpha^7\}$ or $\{\alpha^{11}, \quad \alpha^{12}, \quad \alpha^{13}\}$

$\qquad \to$ three consecutive roots

$A = 19 \cdot 7 \to$ one strong factor $\to$ one root only

CONJECTURE   (Chien–Hong [3])

If A had d consecutive roots covered by the strong factors of A, then the cyclic AN codes have distance greater than or equal to $d + 1$.

This conjecture has been verified by a computer search for all $n \le 36$, thereby lending strong credence to its validity. However, the actual $d_{\min}$ is in most cases greater than the one obtained by this conjecture and thus this lower bound for distance is rather a loose bound. Therefore, Chein and Hong suggested selecting weak factors A_i of A as follows. Let A_i, A_j be factors of A (i and j each dividing n), but $A_i A_j$ not a factor. Also let A_i be a proper divisor of A_j. Then A_j is selected as a weak factor. By using the weak factors and their consecutive roots, a much tighter bound for distance (which in most cases is the exact $d_{\min}$) is obtained. Unfortunately in two cases, for $n \le 36$, this bound does not hold, and therefore some further work is required on this subject.

Special case bound   (Hartman and Tzeng)

Hartman and Tzeng [4] present an upper bound on $d_{\min}$ of cyclic AN codes of $n = n_1 n_2$. This bound is rather tight and therefore provides a rather good estimate of $d_{\min}$.

THEOREM 8.1   (Hartman)

Let a cyclic AN code be of length $n = cn_1$ for some integer $c > 1$. If $(A, r^c - 1) = 1$, then $d_{\min} \leq n_1$.

Proof

The code is in radix r, and we observe that

$$r^n - 1 = r^{cn_1} - 1 = (r^c - 1)(r^{c(n_1-1)} + r^{c(n_1-2)} + \cdots + r^c + 1) \quad (8.7)$$

Since $A \mid (r^n - 1)$ and $(A, r^c - 1) = 1$ we have that A divides $I = r^{c(n_1-1)} + r^{c(n_1-2)} + \cdots + r^c + 1$. Since the arithmetic weight of I is n_1, the arithmetic weight of the AN code is less than or equal to n_1, and the theorem is proved.

As an example consider the AN code $A = 31 \cdot 151$ of length $n = 15 = 5 \cdot 3$. Since $(A, 2^3 - 1) = 1$, from Theorem 8.1 we have $d_{\min} \leq d_u = 5$ (d_u is the upper bound for $d_{\min}$). Hartman provides a table of actual $d_{\min}$ and d_u for several codes and demonstrates that d_u is a reasonably good bound in most cases.

Other results on $d_{\min}$

The following result is due to Erosh and Erosh [5] and is reported by Chien et al. [6].

THEOREM 8.2

The $d_{\min}$ of a cyclic AN code generated by $(2^{n_1 n_2} - 1)/(2^{n_1} - 1)$ is n_2.

Proof

$A = (2^{n_1 n_2} - 1)/(2^{n_1} - 1)$ is of the form

$$A = 2^{n_1(n_2 - 1)} + 2^{n_1(n_2 - 2)} + \cdots + 2^{n_1} + 1 \qquad (8.8)$$

Clearly for $n_1 > 1$, $W(A) = n_2$. For the cyclic AN code $0 < N < 2^{n_1} - 1$, N is a binary integer of n_1 bits or less, and from (8.8) its codeword AN is a repetition of that binary form of N a total n_2 times. It is clear that the minimum weight (and distance) of the code is n_2, and that the code corrects errors of weight up to $[(n_2 - 1)/2]$.   Q.E.D.

Chien *et al.* [6] give the following.

THEOREM 8.3

For the cyclic AN code generated by

$$A = (2^{n_1 n_2} - 1)/(2^{n_1} - 1)(2^{n_2} - 1)$$

where $(n_1, n_2) = 1$ and $n_2 > n_1$, the minimum distance equals n_1.

The proof of this theorem is rather long and requires a few lemmas. Since the proof is not really required for the understanding of the rest of this section it is omitted. The interested reader can find the proof in the work of Chien *et al.* [6].

More on root–distance relationships

Recently Chien *et al.* [6] reported a class of multiple-error-correcting codes that possess a root–distance relationship similar to that of the BCH bound or Chien–Hong conjecture. These codes are cyclic codes of length P_1P_2 and are called P_1P_2-codes. The generators of P_1P_2-codes are the following cyclotomatic factors:

$$A_1 = Q_1(2) = 1$$
$$A_2 = Q_{P_1}(2) = 2^{P_1} - 1$$
$$A_3 = Q_{P_2}(2) = 2^{P_2} - 1$$
$$A_4 = Q_{P_1}(2)Q_{P_2}(2) = (2^{P_1} - 1)(2^{P_2} - 1)$$
$$A_5 = Q_{P_1}(2)Q_{P_1P_2}(2) = (2^{P_1P_2} - 1)/(2^{P_2} - 1)$$
$$A_6 = Q_{P_2}(2)Q_{P_1P_2}(2) = (2^{P_1P_2} - 1)/(2^{P_1} - 1)$$
$$A_7 = Q_{P_1P_2}(2) = (2^{P_1P_2} - 1)/(2^{P_1} - 1)(2^{P_2} - 1)$$

Table 8.2 lists the number of successive powers of u, where u is the (P_1P_2)th root of unity, which belong to A_i and the actual $d_{\min}$ of the code. From the results given in Table 8.2, we see that for all possible positive-factor generators of a P_1P_2-code (except the case $P_1P_2 = 6$), the minimum distance of the code is at least one greater than the number of consecutive roots contained in the generator—the same root–distance relationship as that given by the BCH theorem for polynomial codes. As for the case of $P_1P_2 = 6$, the following argument can be given.

Table 8.2

Generator	Number of consecutive roots	Minimum distance
$A_2 = Q_{P_1}(2)$	1	2
$A_3 = Q_{P_2}(2)$	1	2
$A_4 = Q_{P_1}(2)Q_{P_2}(2)$		
If $P_1 = 2$, $P_2 \neq 3$	3	4
If P_1, P_2 are odd	2	4
If $P_1P_2 = 6$	3	3
$A_5 = Q_{P_1}(2)Q_{P_1P_2}(2)$	$P_1 - 1$	P_1
$A_6 = Q_{P_2}(2)Q_{P_1P_2}(2)$	$P_2 - 1$	P_2
$A_7 = Q_{P_1P_2}(2)$	$P_1 - 1$	P_1

Since $2^6 - 1 = Q_1(2)Q_2(2)Q_3(2)Q_6(2)$, where $Q_1(2) = 1$, $Q_2(2) = 3$, $Q_3(2) = 7$, and $Q_6(2) = 3$, the generator A_4 is 21. The code generated by A_4 is of distance 3, which equals the number of consecutive roots contained in $Q_2(x)Q_3(x)$. However, the generator $A_4 = 3 \cdot 7$ can also be considered as the product of $Q_3(2)$ and $Q_6(2)$ since $Q_6(2) = Q_2(2) = 3$. The number of consecutive roots in $Q_3(x)Q_6(x)$ is only 2, which is indeed one less than the minimum distance of the code. The situation that a positive-factor generator can be expressed as different products of cyclotomic factors can only happen when $P_1P_2 = 6$, since Dickson [7] has shown that $Q_i(2) \neq Q_j(2)$ for all $i \neq j$, except $i = 2$ and $j = 6$.

8.4 MAJORITY DECODABLE ARITHMETIC CODES

This class of multiple-error-correcting codes was discovered by Chien *et al.* [6].

If the block length of an arithmetic code is a composite number, then the length n can be expressed as $n = n_1 \cdot n_2$, as shown before. If the generator A of the code is chosen to be

$$A = (2^n - 1)/(2^{n_2} - 1) = 2^{n_2(n_1 - 1)} + 2^{n_2(n_1 - 2)} + \cdots + 2^{n_2} + 1 \quad (8.9)$$

then any codeword V is of the form AN, where $N = 0, 1, 2, \ldots, 2^{n_2} - 2$. We divide the n digits of a codeword into n_1 blocks; then each block is of length n_2. Since N is less than $2^{n_2} - 1$, we have enough positions to express any N in its binary form at each block. Therefore, the binary form of any codeword AN has the property that all its n_1 blocks are identical. It is clear that the minimum distance of the code is n_1, and the code corrects errors of weight up to $(n_1 - 1)/2$. Suppose a correct codeword is changed by an error E, and the result is R; i.e., $R = AN + E$, where $W(E) \leq (n_1 - 1)/2$. To decode, the decoder wants to find AN or E from the result R. We now prove the following lemma and theorems which yield the majority decoding scheme:

LEMMA 8.4

A codeword is a zero word if and only if the arithmetic weight of the result R is less than or equal to $(n_1 - 1)/2$.

The proof of the lemma is obvious and hence omitted. A single arithmetic error may cause several digits to be in error because of the influence of carry or borrow. Sometimes an error may cause a carry process which changes the codeword into an integer greater than $2^n - 1$. In a modulo $2^n - 1$ adder (i.e., 1's complement arithmetic), we have only n positions for the binary representation of the result R. Hence, the carry process may go beyond the position 2^{n-1} and then change cyclically the positions $2^0, 2^1, 2^2, \ldots$, and so on. However, any carry process will stop whenever a zero digit is reached. Similarly, any borrow process will stop whenever a digit 1 is reached. Since the generator is of the form shown in (8.9), it is clear that any nonzero codeword can have at most $n_2 - 1$ consecutive 1's or 0's in its binary. Starting at the erroneous position, any carry or borrow process caused by a single error cannot propagate more than $n_2 - 1$ positions. In other words, the carry or borrow process caused by a single error can change at most n_2 consecutive digits of a nonzero word. (We note that 2^{n-1} and 2^0 are considered to be consecutive.) Thus, we have proved the following theorem.

THEOREM 8.5

A single arithmetic error at any position of a nonzero word can change at most n_2 consecutive digits of this codeword.

We can express a correctable error pattern E in its NAF as

$$E = e_1 2^{a_1} + e_2 2^{a_2} + \cdots + e_t 2^{a_t}, \qquad \text{where} \qquad e_i = \pm 1;$$
$$t \le (n_1 - 1)/2$$

Among the n digits of R, let us consider the set of digits with positions $2^k, 2^{n_2+k}, 2^{2n_2+k}, \ldots, 2^{(1-1)n_2+k}$, where $0 \le k \le n_2 - 1$. All these positions are $n_2 + 1$ digits apart from one another, so no two of them can be changed by a single error. Since the weight of error E is t, at most t of the positions 2^k, 2^{n_2+k}, 2^{2n_2+k}, ..., $2^{(n_1-1)n_2+k}$ can be altered by E, the rest of the positions remaining unchanged. But t is less than or equal to $(n_1 - 1)/2$, thus we have proved the following theorem.

THEOREM 8.6

If the error is of weight less than or equal to $(n_1 - 1)/2$, then in the binary form of R, the majority of the digits with positions

$$2^k, \quad 2^{n_2+k}, \quad 2^{2n_2+k}, \quad ..., \quad 2^{(n_1-1)n_2+k} \qquad \text{where} \quad 0 \le k \le n_2 - 1$$

remain the same as the correct codeword.

From the lemma and theorems, we now summarize the decoding algorithm for the code generated by $(2^n - 1)/(2^{n_2} - 1)$ as follows:

ALGORITHM 8.1   (One-step majority decoding)

If $W(R) \le (n_1 - 1)/2$, we decode it as a zero word. If $W(R) > (n_1 - 1)/2$, we work on the sets of digits in R with positions $2^k, 2^{n_2+k}, ...,$ $2^{(n_1-1)n_2+k}$ for $k = 0, 1, 2, ..., n_2 - 1$. For each set, we take the majority value of these digits, and form a block of length n_2 with the n_2 digits obtained from each set.

From Section 8.3, it is seen that the generators A_5 and A_6 of a P_1P_2-code are of the same form as $(2^n - 1)/(2^{n_2} - 1)$ for $n = n_1 n_2$, and hence the codes generated by A_5 and A_6 are majority logic decodable in one step.

Chien *et al.* [6] show that the P_1P_2-code generated by $A_7 = (2^{P_1P_2} - 1)/(2^{P_1} - 1)(2^{P_2} - 1)$ is also majority decodable but in two steps, provided $P_2 > 2P_1$, as follows:

ALGORITHM 8.2   (Two-step majority decoding)

Step 1

Multiply the result R to be decoded by $(2^{P_2} - 1)$ and obtain the residue of $R(2^{P_2} - 1)$ modulo $M(M = 2^{P_1P_2} - 1)$. Divide $|R(2^{P_2} - 1)|_M$ in P_2 blocks and make a majority decision to obtain E'.

Step 2

Divide $E'/(2^{P_2} - 1)$ into P_1 blocks and apply a majority decision to obtain the actual error E.

The two-step majority logic decoding scheme has also been generalized to L-step majority decoding by using composite $n = \prod_{i=1}^{L} P_i^{S_i}$, where $P_i > 2P_{i-1}$ for $i = 2, 3, \ldots, L$. For proof of this and examples the reader is advised to refer to the work of Chien *et al.* [6].

The majority decoding scheme is extremely attractive when compared with the decoding schemes of Laste and Tsao-Wu [8], Hong [9], and Monteiro and Rao [10]. These schemes use permutation (or cycling) of residues until a canonical (or standard) form is reached and then search (or table look-up) is employed, although the search is considerably less than that required by a brute force approach. However, the majority decodable codes have a very low information rate and therefore are least attractive from the cost or redundancy point of view.

8.5 SELF-CHECKING PROCESSORS

In earlier sections, we discussed the error detection and correction properties of arithmetic codes and their implementation schemes. The requirement of ultrareliable spacecraft computers for long missions to the outer planets of the solar system has led to the development and design of a number of self-checking (self-testing) processor organizations. Among these we summarize here: (1) a residue checker organization for an arithmetic and logic unit (ALU) studied and reported by the electrical engineering department of the University of Maryland under a NASA grant; (2) the self-testing and repair (STAR) computer concepts developed and reported by Avizienis and his associates at the jet propulsion laboratory.

Residue checker for arithmetic and logic unit (ALU)

The separate residue $[N, |N|_b]$ code is shown to be closed under the operations of ADD, SUBTRACT, and CYCLE when the check base b divides m, the range of information. The code is not closed under other operations, such as SHIFT, but by suitable corrections applied to the checker from the processor, the code can be maintained so that error checking of these operations is made possible. The $[N, |N|_b]$ code is not closed under logical operations, AND, OR, EXOR. The parity-based codes, such as the Hamming code [1], is closed under EXOR operation, but not under AND, OR. These parity codes are not useful for arithmetic operations either, although some work favoring a parity code for addition is reported [11, 12]. Monteiro and Rao [13] have shown that a separate residue checker using a check base of the form $2^c - 1$ can also be used to check logical operations very efficiently, and they have also given details of design and analysis of a checker for arithmetic and logical operations. Their scheme thus eliminates the need for different codes for logical and arithmetic operations.

The checking scheme maintains a residue check $|R|_b$ in the checker corresponding to the result R obtained in the processor for all operations; while the ALU uses AND, OR, EXOR gates to obtain R, the checker uses only AND, ADD, CYCLE operations to obtain $|R|_b$. The increase in hardware cost of the checker due to logical operations is minimal and the overall hardware cost of the checker (redundancy) is shown to be less than 50% of the cost of the processor. The logical and arithmetic operations are concurrently checked without loss of speed. For details the reader is advised to see Monteiro and Rao [13].

JPL STAR computer

Avizienis and his associates have conducted a systematic, practically oriented study on the concepts of the design of the self-testing and repair computer (STAR) [14, 15]. Their results are summarized here in brief.

The concept of concurrent error diagnosis and error recovery can be implemented by (1) triplication and majority vote-taking; (2) duplication of all essential modules (or functional blocks) and comparison of their outputs, followed usually by a software diagnosis and switching of spare modules; (3) arithmetic coding of the operands in such a way that the results are decoded for error detection–correction. The last method is the least expensive in terms of hardware and power requirements but can provide only limited protection at this time. Avizienis has developed algorithms for checking arithmetic operations which include SHIFT, range extension, contraction, and so on. The effectiveness in terms of fault coverage and the implementation facility of AN codes and separate codes have been studied in detail. Check bases of the form $2^c - 1$ are shown to offer a "low-cost checking algorithm" [15] in terms of hardware cost and speed of operation.

Avizienis [15] has also studied the use of the check base $2^c - 1$ for a serial, byte-organized arithmetic processor. An important result of this study is that the checking algorithm is very efficient when the byte length coincides with c of the check base $2^c - 1$, and that the maximal word length in this case is given by $c(2^c - 2)$. This result also follows from Theorem 7.5.

Further design considerations on fault recovery and spare-module switching for the JPL STAR have been investigated and reported by Avizienis [15].

Conclusion

Arithmetic codes, their categories, their error detection and correction properties, and implementation schemes have been discussed here. The separate codes, particularly the biresidue code organization using check bases of the form $2^c - 1$, are found to be very attractive for error correction of all arithmetic operations. The separate arithmetic codes can also be applied to check logical operations [13], as well as memory operations. This capability makes them the most important code among all error codes. However, there are a number of important considerations to be borne in mind in choosing the code or codes for application. These are discussed below.

The biresidue code using check bases of the form $2^c - 1$ has a lower information rate (therefore greater redundancy) than a Hamming distance 3 (or 4) code, and is therefore expensive when large main memories are required. The biresidue codes using check bases not of the form $2^c - 1$ have relatively higher information rates, somewhat comparable to a Hamming distance 4 code, and therefore save memory hardware. However, their implementation is complex, and therefore expensive. Alternatively, one may suggest using a Hamming code for memory words (or memory operations) and a biresidue code (using bases of the form $2^c - 1$) for arithmetic and logical operations. But this will have the disadvantage of (1) too many encoders and decoders, i.e., hardware expense, and (2) slower operation or longer operation cycle due to reencoding and decoding for the different codes used. Since these disadvantages are severe, one tends to favor using one code, a biresidue code (check bases not of the form $2^c - 1$), for all operations, and to develop efficient, workable algorithms for its implementation.

REFERENCES

1. W. W. Peterson and E. J. Weldon, Jr., "Error Correcting Codes." MIT Press, Cambridge, Massachusetts, 1972.
2. E. R. Berlekamp, "Algebraic Coding Theory." McGraw-Hill, New York, 1968.
3. R. T. Chien and S. J. Hong, "On Root-distance Relation for Arithmetic Codes," Rep. R-440. Coordinated Sci. Lab., Univ. of Illinois, Urbana, October 1969.
4. C. R. P. Hartman and K. K. Tzeng, A Bound for Arithmetic Codes of Composite Length, *IEEE Trans. Information Theory* **IT-18**, 308 (March 1972).
5. I. L. Erosh and E. L. Erosh, Arithmetic Codes with Correction of Multiple Errors, *Problemy Peredaci Informacii* **3**, 72–80 (1968).
6. R. T. Chien, C. K. Liu, and C. L. Chen, "On a Class of Majority Decodable Arithmetic Codes," Rep. Coordinated Sci. Lab., Univ. of Illinois, Urbana 1972.
7. L. E. Dickson, On the Cyclotomic Function, *Amer. Math. Monthly* **12**, 86–89 (1905).

8. E. R. Laste and N. T. Tsao-Wu, The Decoding of Arithmetic Cyclic Codes by Permutation of Residues. *Proc. Annu. Conf. Information Sci. and Systems, 3rd, Princeton, New Jersey, March 1969.*

9. S. J. Hong, "On Bounds and Implementation of Arithmetic Codes," Rep. R-437. Coordinated Sci. Lab., Univ. of Illinois, Urbana, October 1969.

10. P. M. Monteiro and T. R. N. Rao, Multi-Residue Codes for Double Error Correction, *Proc. IEEE-TCCA Symp. Comput. Arithmetic,* **vol. 1,** *Univ. of Maryland, College Park, May 1972.*

11. D. K. Pradhan and S. M. Reddy, Fault-Tolerant Adders, *Proc. Internat. Symp. Fault-Tolerant Comput., Boston, Massachusetts, June 1972.*

12. G. G. Langdon, Jr. and C. K. Tang, Concurrent Error Detection for Group-Carry-Look-Ahead Binary Adders, *IBM J. Res. Develop.* **14,** (September 1970).

13. P. M. Monteiro and T. R. N. Rao, A Residue Checker for Arithmetic and Logical Operations, *Proc. Internat. Symp. Fault-Tolerant Comput., Boston, Massachusetts, June 1972.*

14. A. Avizienis, G. C. Gilley, F. P. Mathur, D. A. Runnels, J. A. Rohr, and D. K. Rubin, The STAR (Self-Testing and Repair) Computer: An Investigation of the Theory and Practice of Fault-Tolerant Computer Design. *IEEE Trans. Comput.* **C-20,** 1312–1321 (November 1971).

15. A. Avizienis, Arithmetic Error Codes: Cost and Effectiveness Studies for Application in Digital System Design, *IEEE Trans. Comput.* **C-20,** 1322–1330 (November 1971).

INDEX

Numbers in italics indicate pages on which complete reference information may be found.

ELECTRICAL SCIENCE

A Series of Monographs and Texts

Editors

Henry G. Booker

UNIVERSITY OF CALIFORNIA AT SAN DIEGO
LA JOLLA, CALIFORNIA

Nicholas DeClaris

UNIVERSITY OF MARYLAND
COLLEGE PARK, MARYLAND

C. L. Sheng. Threshold Logic. 1969

Dale M. Grimes. Electromagnetism and Quantum Theory. 1969

Robert O. Harger. Synthetic Aperture Radar Systems: Theory and Design. 1970

M. A. Lampert and P. Mark. Current Injection in Solids. 1970

W. V. T. Rusch and P. D. Potter. Analysis of Reflector Antennas. 1970

Amar Mukhopadhyay. Recent Developments in Switching Theory. 1971

A. D. Whalen. Detection of Signals in Noise. 1971

J. E. Rubio. The Theory of Linear Systems. 1971

Keinosuke Fukunaga. Introduction To Statistical Pattern Recognition. 1972

Jacob Klapper and John T. Frankle. Phase-Locked and Frequency-Feedback Systems: Principles and Techniques. 1972

Kumpati S. Narendra and James H. Taylor. Frequency Domain Criteria for Absolute Stability. 1973

Daniel P. Meyer and Herbert A. Mayer. Radar Target Detection: Handbook of Theory and Practice. 1973

T. R. N. Rao. Error Coding for Arithmetic Processors. 1974

In Preparation

C. A. Desoer and M. Vidyasagar. Feedback Systems: Input-Output Properties

A 4
B 5
C 6
D 7
E 8
F 9
G 0
H 1
I 2
J 3